Nanobioelectrochemistry

Frank N. Crespilho
Editor

Nanobioelectrochemistry

From Implantable Biosensors to Green Power Generation

Springer

Editor
Frank N. Crespilho
Institute of Chemistry of São Carlos (IQSC)
University of São Paulo (USP)
São Carlos
13560-970
Brazil

ISBN 978-3-642-29249-1 ISBN 978-3-642-29250-7 (eBook)
DOI 10.1007/978-3-642-29250-7
Springer Heidelberg New York Dordrecht London

Library of Congress Control Number: 2012940226

Printed on acid-free paper

Springer is part of Springer Science+Business Media (www.springer.com)

Preface

Nanobioelectrochemistry covers the modern aspects of bioelectrochemistry, nanoscience, and materials science. The combination of nanostructured materials and biological molecules enable the development of biodevices capable of detecting specific substances. Furthermore, by using the bioelectrochemistry approach, the interaction between bio-system and nanostructured materials can be studied at molecular level, whcre several mechanisms of molecular behavior are elucidated from redox reactions. The combination of biological molecules and novel nanomaterials components is of great importance in the processes of developing new nanoscale devices for future biological, medical, and electronic applications. This book describes some of the different electrochemical techniques that can be used to study new strategies for patterning electrode surfaces with biomolecules and biomimetic systems. Also, it focuses on how nanomaterials can be used in combination with biological catalysts in fuel cells for the green power generation. By bringing together these different aspects of nanobioelectrochemistry, this book provides a valuable source of information for many students and scientists.

Chapters from Implantable Biosensors to Green Power Generation

This book provides a comprehensive compilation of seven chapters, with important contributions of several authors. Chapter 1 reviews the recent research in using nanoparticle labels and multiplexed detection in protein immunosensors. This chapter summarizes recent progress in development of ultrasensitive electrochemical devices to measure cancer biomarker proteins, with emphasis on the use of nanoparticles and nanostructured sensors aimed for use in clinical cancer diagnostics. Based on recent strategies focused on nanomaterials for electrochemical biosensors development, Chap. 2 discusses the development of new methodologies for biomolecules immobilization; including the utilization of several biological molecules such as enzymes, nucleotides, antigens, DNA, aminoacids, and many others for biosensing. The utilization of these biological molecules in conjunction with nanostructured materials opens the possibility to develop several types of biosensors such as nanostructured and miniaturized devices and implantable biosensors for real-time monitoring. Also, nanomaterials, such as carbon nanotubes, seem to be the most appropriate electrical host matrix in biofuel cells due to their bio-compability, high conductivity, high specific surface, and ability to electrically connect many redox enzymes. The latter, is the focus of Chap. 3, which also shows that biofuel cells attract more and more attention as green and non-polluting energy source for, in general, mobile and implantable devices. Within this research topic discussed in this chapter, nanostructured materials prevail due to their higher efficiency, energy yields, and the possibility to construct miniaturized devices. This can also lead to development and applicability of implantable devices, when biosensing has benefitted enormously from the development of field-effect transistor (FET) sensor platforms, not only due to the design of specific FET architectures, but also because nanotechnological materials and techniques may be used to obtain gate platforms with tailored surfaces and functionalities. This topic is presented in Chap. 4, in which is shown the crucial points for improving the efficiency of biomolecules immobilization, leading to higher protein loadings, and as a consequence, better sensitivity and lower limit of detection. Another advantage is the number of possible architectures leading to distinct devices including ion-sensitive field-effect transistor (ISFET), electrolyte-insulator-semiconductor (EIS), light-addressable potentiometric sensor, extended-gate field-effect transistor (EGFET), and separative extended-gate field-effect transistor (SEGFET), each of which exhibits advantages for specific applications. Also, Chap. 5 show how the supramolecular chemistry strategy is used to map electrochemical phenomena at the nanoscale of low-dimensional highly organized hybrid structures containing several building blocks such as metallic nanoparticles, carbon nanotubes, metallic phthalocyanine, biopolymers, enzymes, and synthetic polymers. The principles of supramolecular chemistry as constitutional dynamic character of the reactions, functional recognition, and self-organization are explored from interaction between biomolecules and several supramolecular architectures in order to modulate the physicochemical properties that arise at molecular level. The developed platforms with high control of these electrochemical properties become interesting devices for sensor and biosensor applications. Chapter 6 illustrates recent developments on surface characterization of DNA and enzyme-based sensors to complement information obtained by electrochemical and impedance techniques. This chapter also shows how AFM imaging is used to characterize different procedures for immobilizing nanoscale double-stranded DNA surface films on carbon electrodes, in which a critical issue is the sensor material and the degree of surface coverage. In this regard, another important technique is the Electrochemical-Surface Plasmon Resonance (ESPR). The combination of SPR and electrochemical methods has become a powerful technique for simultaneous observation of optical and electrochemical properties at substrate/electrolyte interfaces, as shown in the Chap. 7. The fundamental aspects of the electric potential effects on surface plasmons are introduced and the use and applications of this combined electrochemical and optical technique are discussed.

Contents

Chapter 1
Nanoscience-Based Electrochemical Sensors and Arrays for Detection of Cancer Biomarker Proteins

James F. Rusling, Bernard Munge, Naimish P. Sardesai, Ruchika Malhotra and Bhaskara V. Chikkaveeraiah

Abstract Measurement of panels of biomarker proteins in serum, tissue or saliva holds great promise for future cancer diagnostics. Broad implementation of this approach in the clinic requires new, low cost devices for multiplexed protein detection. Advanced nanomaterials coupled with electrochemical detection have provided new opportunities for development of such devices. This chapter reviews recent research in using nanoparticle labels and multiplexed detection in protein immunosensors. It focuses in part on research in our own laboratories on ultra-sensitive protein immunosensors combining nanostructured electrodes with detection particles with up to 500,000 labels that detect as little as 1 fg/mL protein in diluted serum. Our most mature multiple protein detection arrays are multiplexed microfluidic devices with 8-nanostructured sensors utilizing massively labeled magnetic particles or polymers. This approach provides reliable detection for multiple proteins at levels well below 1 pg/mL, and shows by excellent correlation with referee methods. The importance of validating panels of biomarkers for reliable cancer diagnostics is also stressed.

J. F. Rusling (✉) · N. P. Sardesai · R. Malhotra · B. V. Chikkaveeraiah
Department of Chemistry (U-3060), University of Connecticut, 55 North Eagleville Road, Storrs, CT 06269, USA
e-mail: James.Rusling@uconn.edu

J. F. Rusling
Department of Cell Biology, University of Connecticut Health Center, Farmington, CT 06032, USA

J. F. Rusling
Institute of Materials Science, University of Connecticut, 97 North Eagleville Road, Storrs, CT 06269, USA

B. Munge
Department of Chemistry, Salve Regina University, 100 Ochre Point Ave., Newport, RI 02840, USA

F. N. Crespilho (ed.), *Nanobioelectrochemistry*, DOI: 10.1007/978-3-642-29250-7_1,

1.1 Introduction

The concentrations of many cancer-related proteins increase in blood during the onset of the disease. Measurements of these proteins hold great promise to detect specific cancers and to monitor their treatment [1–3]. While the possibility to clinically assess concentrations of panels of cancer biomarker proteins has created great interest [4–7], broad implementation of such strategies has lagged somewhat behind research because of the lack of technically reliable, inexpensive devices to measure multiple proteins in patient samples in a clinical setting [2]. On the other hand, such devices have enormous potential to improve cancer diagnostics. Among possible methodologies, electrochemical approaches enhanced by nanomaterials offer the potential for high sensitivity, high selectivity, low cost, and instrumental simplicity [2, 8, 9]. This chapter reviews progress in research aimed toward electrochemical detection of multiple biomarker proteins, focusing on sensors that derive signals from active oxidation and reduction processes. There have been parallel developments in nanowire transistors for proteins that we have not included here [10, 11].

The next section discusses the nature and significance of biomarker proteins for cancer, followed by a section reviewing the use of nanoparticles in sensors and detection protocols. The section following discusses the combination of nanoscience-assisted sensing with microfluidics for multiplexed protein detection. We end the chapter with an overview and comments on the future of cancer diagnostics based on biomarker detection.

1.2 Biomarker Proteins and Cancer

The US National Institutes of Health defines biomarkers as "molecules that can be objectively measured and evaluated as indicators of normal or disease processes and pharmacologic responses to therapeutic intervention" [12]. A broader definition of biomarkers for cancer consist of any measurable or observable factors in a patient that indicate normal or disease-related biological processes or responses to therapy [13, 14]. These can include physical symptoms, mutated DNAs and RNAs, secreted proteins, cell death or proliferation, and serum concentrations of small molecules such as glucose or cholesterol. In this chapter, we focus on emerging nanoscience-based electrochemical methods to detect levels of *proteins* as biomarkers that can be used for detection and monitoring cancer [2, 6, 15].

Many proteins are overexpressed and secreted into the blood beginning at very early stages of developing cancers. Serum levels of these proteins can indicate cancer and guide therapy even before the onset of detectable tumors. Biomarker proteins are often specific to several types of cancer, and panels of such proteins promise a more much reliable assessment of patient status than single biomarkers [2, 5, 8, 16, 17]. The most famous clinically used single biomarker protein is prostate specific antigen (PSA), a prostate cancer biomarker. The PSA serum test has an insufficient positive predictive value of ~70 % [18], leading to false positives and unnecessary treatment.

A number of technologies exist or are being developed for protein detection [2, 6, 15, 19]. Many utilize nanomaterials such as quantum dots, gold nanoparticles, carbon nanotubes and magnetic particles to enhance sensitivity [20]. Low detection limits achieved by using nanomaterials can facilitate early cancer detection and accurate prognosis. Devices for clinical or point-of-care (POC) detection of panels of proteins must be sensitive, multiplexed, accurate, and reasonably priced. POC requirements are more demanding, and include speed, automated sample preparation, low cost, and technical simplicity. These requirements have not been fully met by any available methodology to date. Ideally, the device should be able to accurately measure both normal and elevated serum levels of proteins. Concentrations in serum that need to be measured may be in the sub-pg mL^{-1} to high ng mL^{-1} ranges for different proteins. Potential interferences include the many thousands of proteins present in serum, some at relatively high levels [2, 5]. In addition to development of the devices, appropriate panels of proteins for specific cancers will need to be validated for accuracy with patient samples. Studies will also be needed to establish the diagnostic value of specific biomarker panels [21, 22], preferably using the new clinical measurement technologies.

1.3 Nanomaterials in Protein Sensing Devices

1.3.1 Nanomaterials in Electrochemical Immunoassays

Electrochemical methodology for protein detection has been provided exciting new opportunities by the revolution in nanotechnology. In particular, nanostructured electrodes, nanoparticle labels, and magnetic nanoparticles for analyte manipulation have featured heavily in strategies for high sensitivity protein detection [2, 6, 8, 9, 15]. Most of these approaches have adapted the sandwich immunoassay from enzyme-linked immunosorbent assays (ELISA), which have served as workhorse methods for clinical protein determinations. Although protein detection limits (DL) in classic ELISA approach 1 pg mL^{-1} [23], the method has limitations in analysis time, sample size, equipment cost, and measuring collections of proteins.

An ELISA-like sandwich assay is illustrated for a hypothetical array format in Fig. 1.1. Spots in the array are shown on an underlying nanoparticle bed, and may contain capture antibodies or aptamers on the spots to capture analyte proteins from the sample. After washing with detergent-protein solutions designed to block non-specific binding (NSB), a labeled secondary antibody is added to bind to captured analyte proteins. Enzyme labels catalyze conversion of an added chemical substrate to produce a colored product that is usually measured with an optical

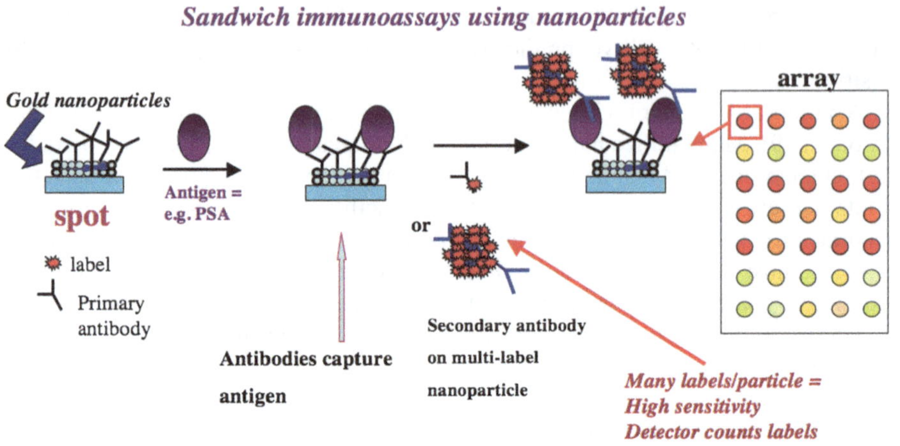

Fig. 1.1 ELISA-like immunoarray strategy to detect proteins (PSA=prostate specific antigen) demonstrating some uses of nanoparticles. Gold nanoparticles on the spot areas are linked to primary antibodies that capture the protein analytes. After washing, a labeled secondary antibody, or as illustrated here a multi-labeled nanoparticle with attached secondary antibody, is added. This detection particle binds selectively to the captured analyte molecules. After additional washing with blocking agents to remove non-specific binding of the labeled species, electrical or optical detection is used to "count" the number of bound labels that is proportional to protein analyte concentration

plate reader. Multiple labels provide higher sensitivity [2, 23, 24], and detection could also involve amperometry, voltammetry, impedance or other electrochemical methods in different formats.

Heinemann et al. pioneered electrochemical immunoassays prior to the nanoparticle era [25]. His team's systems involve sandwich immunoassays using the enzyme label alkaline phosphatase which produces electroactive products that are transported by a chromatographic or fluidic system to an electrode detector [26, 27]. Recent advances have interfaced this approach into microfluidic devices [28]. Interdigitated electrodes have provided the highest sensitivity [29].

Self-contained, single analyte electrochemical immunosensors [30–34] that feature antibodies (Ab) attached to the sensor surface have also been developed. This approach has the advantage that protein analyte capture, binding of the enzyme-labeled secondary antibody, and detection are all done on the sensor surface. Alkaline phosphatase, glucose oxidase and horseradish peroxidase (HRP) have been used as enzyme labels along with suitable substrates.

Multiplexing has also been achieved with electrochemical immunosensors. Separation of iridium oxide electrodes by 2.5 mm in arrays to eliminate cross-talk enabled simultaneous electrochemical immunoassays using alkaline phosphatase-labeled Ab_2 and detection of product hydroquinone giving DLs ~ 1 ng mL^{-1} for goat IgG, mouse IgG, and cancer biomarkers carcinoembryonic antigen (CEA) and α-fetoprotein (AFP) [35]. An 8-electrode array was developed for detection [36] of

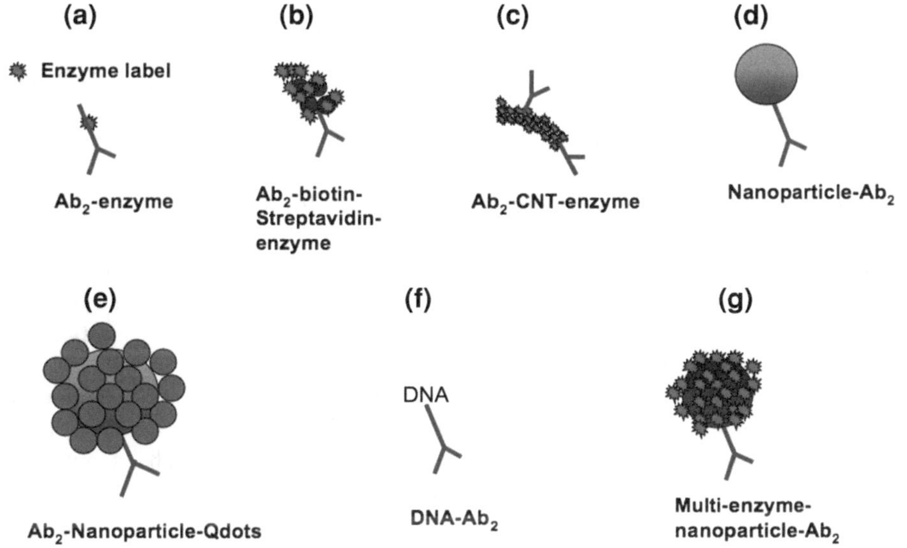

Fig. 1.2 Amplification particles for electrochemical immunosensors featuring nanoparticles or other moieties attached to secondary antibody Ab_2

goat IgG, mouse IgG, human IgG, and chicken IgY with DLs of $\sim$3 ng mL^{-1}. Eight-electrode iridium oxide arrays in each well of a 12-well plate were used to simultaneously measure [37] cancer biomarkers AFP, ferritin, CEA, hCG-*b*, CA 15-3, CA 125, and CA 19-9 with DLs of $\sim$2 ng mL^{-1}. The method showed good correlation with ELISA for proteins in standard serum.

The advent of simple, reliable methodology for nanoparticle fabrication has led to new, ultrasensitive approaches to electrochemical protein detection [2, 8–10]. Cancer detection and monitoring using protein biomarker panels in serum requires detection limits below that of the normal patient concentrations and sensitivity for all the protein biomarkers at normal and elevated levels. Detection limits below pg mL^{-1} levels and good sensitivity up to hundreds of ng mL^{-1} will be necessary. Strategies using secondary antibody (Ab_2)-nanoparticle bioconjugates in sandwich immunosensors have included dissolvable nanoparticles labels leading to electroactive ions, Ab_2-nanoparticles with thousands of enzyme labels (Fig. 1.2), and Ab_2-nanoparticles with multiple redox probes [38–43]. High sensitivity is achieved in these approaches by providing a large number of signal generating events for each protein bound onto the sensor.

Also, nanostructured electrode surfaces can provide an additional sensitivity boost, both by enabling the attachment of a large number of capture antibodies on the sensor surface [2, 44], and by allowing better access of protein analytes to these antibodies [45]. Nanostructured surfaces for immunosensors have been made using films of carbon nanotubes [10, 43] or gold nanoparticles [2, 46], or by highly nanostructured microsensor surfaces made by electrodepositing gold [45, 47].

In recent applications, Cai et. al. [48] used arrays of vertically aligned carbon nanotube tips with an imprinted non-conducting polymer coating of polyphenol to detect ferritin and human papilloma virus (HPV) biomarker E7 protein using electrochemical impedance spectroscopy for a DL of 10 pg mL^{-1} for ferritin. Osakai et. al. [49] reported label-free voltammetric detection of cytochrome c, lysozyme, myoglobin, and α-lactalbumin at a polarized oil/water interface using anionic surfactants to co-absorb with proteins at the oil/water interface. Genc et. al. [50] reported an amperometric immunosensor utilizing enzyme encapsulated thermosensitive liposomes for detection of carcinoembryonic antigen (CEA). Bioconjugation using N-succinimidyl-S-acetylthioacetate/sulfosuccinimidyl 4-(N-maleimidomethyl) cyclohexane-1-1-carboxylate to link an anti-CEA antibody to liposomes yielded a DL of 11 pg mL^{-1}.

Numerous carbon nanotube (CNT) sensors have been used for detection of proteins. CNTs are commonly functionalized by using acids to shorten the lengths and add terminal carboxylate groups. Additional chemistry, such as antibody attachment, can be done on the functionalized ends. Zhao et. al. [51] reviewed non-covalent functionalization of CNTs, and wrapping with polymers or DNA to tune the electrochemical properties of the sensor. Jacobs et. al. [52] reviewed CNT-based sensors for detection of a variety of proteins using amperometry, voltammetry, and impedance. These new technologies for biosensors face the serious challenges of use in more realistic biological monitoring experiments, such as in serum, blood, saliva or tissue.

We end this section by addressing a critical issue in any immunoassay, minimization of non-specific binding (NSB). There are two types to be inhibited, (1) NSB of any molecule in the sample that interferes with the assay, and (2) NSB of labeled-Ab_2 bound to non-antigen sites on the sensor. In label-free methods such as impedance, NSB tends to be more serious since any biomolecule that binds to the sensor can contribute to the signal. In labeled methods, bound, labeled-Ab_2 will give a signal even if not bound to the capture antibody, but this signal is not proportional to analyte concentration. NSB can increase detection limits (DL) and degrade sensitivity, but can be minimized by washing with blocking solutions of bovine serum albumin or casein containing nonionic detergents such as Tween-20. Derivatizing the sensor surface with the appropriate chemistry may also decrease NSB, with one of the most effective surfaces featuring polyethylene glycol (PEG) moieties [34, 39] Optimizing an NSB blocking protocol for a specific assay is often a trial and error process.

1.3.2 Nanoparticles as Labels in Immunoassays

Nanoparticle labels in sandwich immunoassays were first used by Delequaire et al. [53]. After the antibodies capture the analyte proteins, Ab_2-nanoparticle bioconjugates bind to them. Then, the nanoparticles are dissolved in acid to produce a large number of electroactive metal ions. Using gold nanoparticle-Ab_2 labels, they detected gold ions released after acid dissolution by using anodic stripping voltammetry to obtain a 3 pM DL for IgG in buffer. Wang et al. developed ways to enhance sensitivity even further [38]. Strategies included magnetic accumulation of gold nanoparticles and their use to catalyze precipitation of Ag. These approaches produce large concentrations of electrochemically detectable metal ions for measurement by stripping analysis. For example, Ag-deposition provided a 0.5 ng mL^{-1} (22 pM) DL for cardiac troponin I [54]. Multiple gold nanoparticles have been attached to larger Au spheres and used for Ag-deposition enhancement [38]. Magnetic particles have been equipped with CdS quantum dots (Qdots), then collected magnetically and dissolved for electrochemical stripping detection of Cd, which can be further enhanced by Cd-deposition [38]. Ag-deposition was used in high sensitivity conductivity immunoassays of human IgG in buffer [55]. Other multilabel strategies include loading Ab_2-nanoparticles or Ab_2-polymer beads with electroactive labels such as ferrocene derivatives, and releasing these labels for electrochemical detection [38, 39, 42].

Multiplexed protein detection using the above approaches has also been developed [38]. One approach is to use "bar code" labeling secondary antibodies with distinct nanoparticles with easily detectable electrochemical characteristics, e.g. different dissolvable metals or quantum dots (Qdots) that can be dissolved to give ions with different reduction potentials.

For example, zinc sulfide, copper sulfide, cadmium sulfide, and lead sulfide Qdots were attached to four different secondary antibodies to detect four different proteins [56]. The four different Qdots were dissolved to yield four different metal ions, each associated with a different protein. These were measured by stripping voltammetry after dissolution of the particles following the binding steps. Multiple metal striped rods, spheres or alloy rods were also used for multiplexing. The rods were capped with a gold end for attachment to Ab_2. Upon dissolution, these materials give a series of metal stripping peaks whose peak potentials and relative intensities are associated with individual analyte proteins [38]. Such "bar code" labels have the potential to determine many proteins in patient samples, but this has yet to be reported.

Label-free impedance immunosensors have been developed, but in general these methods may require additional amplification to improve sensitivity [57, 58]. Nevertheless, a capacitance method using a ferri/ferrocyanide probe and a potentiostatic step approach gave DL 10 pg mL^{-1} (500 fM) for IL-6 in buffer [59]. Optimization of experimental protocols in flow injection impedance spectroscopy led to sensitivity in the low aM range for interferon-γ in buffer [60]. Sensitivities have been enhanced using metal nanoparticle labels or AuNP labels that catalyze subsequent Ag deposition [57]. These methods may be promising for future point-of-care applications if NSB from non-analyte proteins in the patient samples can be minimized.

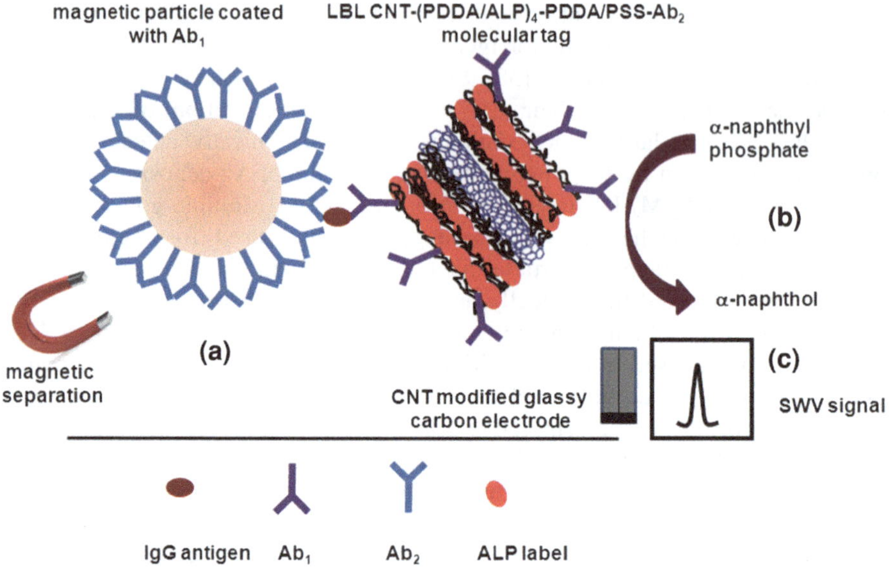

Fig. 1.3 Schematic representation of immunosensor protocol using multilabel CNT; **a** sandwich type immunosensor for the detection of IgG captured on anti-IgG coated magnetic beads, coupled to ALP loaded LBL self-assembled CNT-(PDDA/ALP)4-PDDA/PSS-Ab2 molecular tag; **b** Enzymatic reaction; **c** Electrochemical detection of the enzymatic reaction product, α-naphthol at the CNT modified glassy carbon electrode

1.3.3 Coupling Nanostructured Surfaces with Multilabel Enzyme Detection

Multi-enzyme labeled nanoparticles were first used by Wang et al. for ultrasensitive detection of DNA and proteins [61]. Multiwall carbon nanotubes (MWCNT) were derivatized with thousands of alkaline phosphatase enzymes and secondary antibodies, and used to achieve fM detection of proteins in buffer. MWCNTs also preconcentrated the enzyme reaction product α-napthol by adsorption. Layer-by-layer (LbL) film deposition of alkaline phosphatase (ALP) with oppositely charged polyions on MWCNTs was used to make detection particles and achieve a DL of ~70 aM for IgG in buffer[62] (Fig. 3). The sandwich immunoassay involved Ab_1 on 1 μm magnetic beads, to capture IgG and then a specially designed bioconjugate CNT-$(PDDA/ALP)_4$-PDDA-PSS-Ab_2 particle was made. The electrical signal is generated via biocatalytic reaction of alkaline phosphatase (ALP) in the ALP/LBL/CNT nanoparticle by incubation with naphthyl phosphate. This substrate is converted to α-naphthol, which is detected using a CNT modified glassy carbon electrode. The DL was 2,000 protein molecules (67 aM).

In an alternative approach, a DNA bio-barcode was used for amplified electrochemical detection and coding of proteins [63]. This method employed the oxidation signal from guanine (G) and adenine (A) nucleobases, and included the

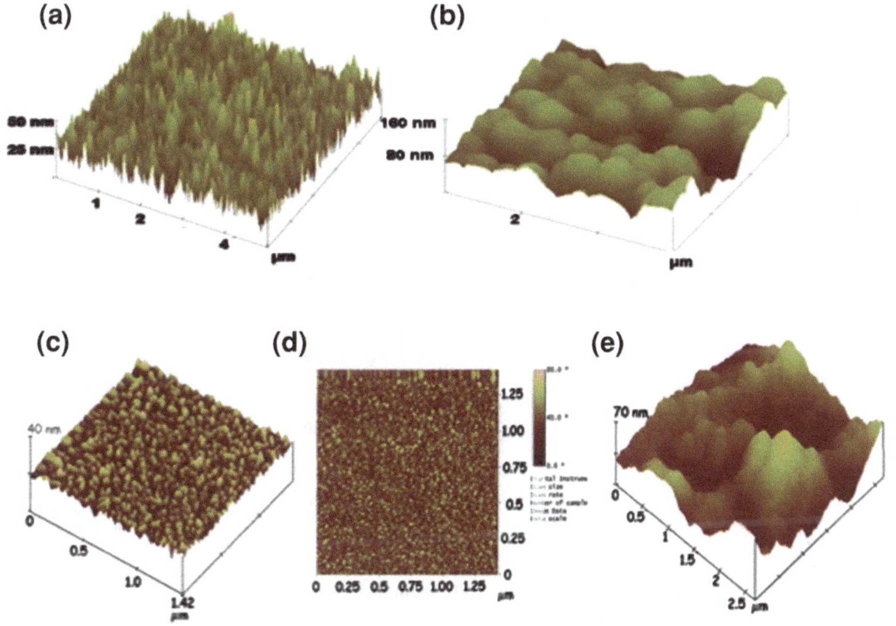

Fig. 1.4 Atomic force microscope images of immunosensor platforms: **a** SWCNT forest on silicon; **b** SWCNT forest coated with chemically attached antibodies; **c** a PDDA/gold nanoparticle (AuNP) bilayer on smooth mica; **d** phase contrast image of the same PDDA/AuNP bilayer; **e** anti-PSA antibodies attached onto carboxylate groups of the AuNP/PDDA bilayer.Reproduced with permission from reference 68 (**a**, **b**), copyright Royal Society of Chemistry, 2005, and reference 46 (**c**–**a**), copyright American Chemical Society 2009.

ability to create oligonucleotide-identifiable bar codes. The sandwich immunoassay was based on two antibodies linked to magnetic beads and DNA-functionalized polystyrene (PS) spheres, followed by the alkaline release of DNA bases that were detected to give DL 2 pg mL^{-1} (13 fM) for mouse IgG. This DNA-based electrochemical method offers promise for the detection of multiple proteins by using identifiable oligonucleotide barcodes in electrochemical immunoassays. Initial assessment of this electrical coding strategy was done using a $dG_{15}A_{10}$ pre-designed oligonucleotide labels that gave distinguishable signals for G and A at different potentials.

Our research team first exploited multi-enzyme labeled nanoparticles in immunosensors for PSA, IL-6, and other prostate cancer biomarkers [2, 10, 38, 43, 64–67] Sensitivity and detection limits were further improved by using nanostructured electrodes featuring densely packed films of oxidatively shortened, upright single wall carbon nanotube (SWCNT) forests [10, 65, 68] or 5 nm glutathione-decorated gold nanoparticles [46]. Both of these surfaces feature large populations of carboxylate groups ready for attachment of large amounts of capture antibodies by amidization[44] AFM images of the films residing on the sensor surfaces illustrate their large surface areas (Fig. 1.4).

Scheme 1.1 Protein detection chemistry using HRP (PFe^{III}) labels

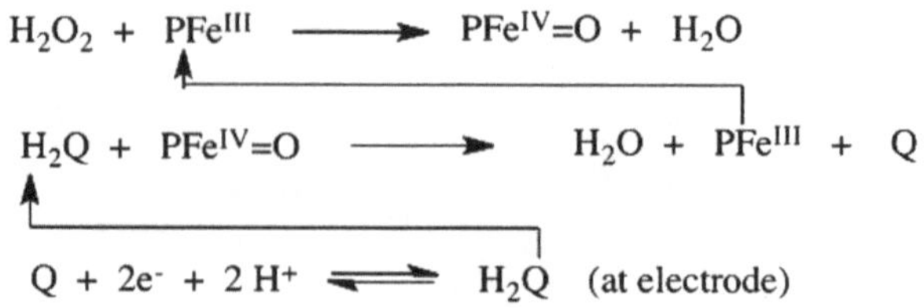

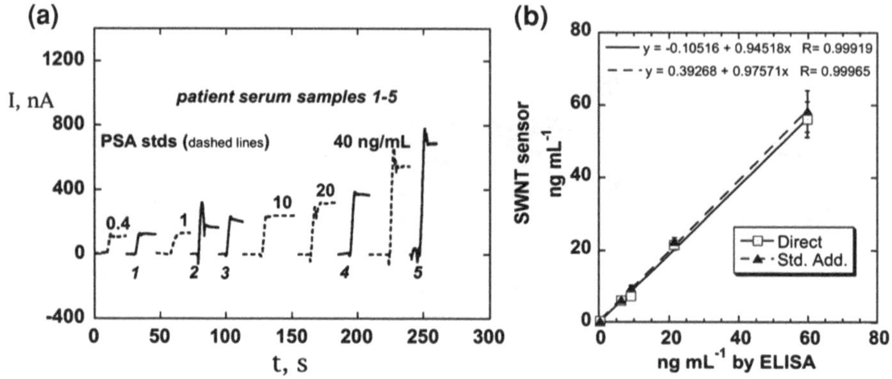

Fig. 1.5 PSA sensor response at –0.3 V and 3000 rpm for human serum samples and PSA standards in calf serum (ng mL^{-1} labeled on curves, dashed lines). SWCNT forest immunosensors were incubated with 10 mL serum for 1.25 h followed by 10 μL 4 pmol mL^{-1} anti-PSA-HRP in 2% BSA and 0.05% Tween-20 for 1.25 h: **a** current after placing electrodes in buffer containing 1 mM hydroquinone mediator, then injecting H_2O_2 to 0.4 mM. Dashed lines are standards in calf serum; solid lines are human serum samples; **b** Correlations of SWNT immunosensor results for human serum samples found by using direct comparison to a calibration curve (□) and by standard addition (◆) against results from ELISA determination (RSD ±10%) for the same samples. Equations shown were found by linear regression. Reproduced with permission from [65], copyright American Chemical Society 2006.

Nanostructured sensors coated with SWCNT forests and AuNP films were used to fabricate sandwich immunoassays for prostate cancer biomarker PSA [10, 43, 46, 64]. As in Fig. 1.2a, conventional secondary antibodies (Ab_2) conjugated with enzyme label HRP were used, as well as carbon nanotubes (CNT) or magnetic particles conjugated with Ab_2 (Fig. 1.2c,g). These heavily labeled detection particles [69] can replace singly-labeled HRP-Ab_2 in immunoassays to greatly enhance sensitivity.

Rotating disk amperometry was used to measure these immunosensor responses using H_2O_2 to activate $HRPFe^{III}$ to a ferryloxyHRP form ($HRPFe^{IV} = O$), and hydroquinone (HQ) to mediate the reduction of $HRPFe^{IV} = O$ (Scheme 1.1). The sensor response is a steady state amperometric current proportional to protein concentration (see Fig. 1.5a).

SWCNT forest immunosensor responses to PSA in calf serum gave a DL as 3X the noise above the zero PSA control of 4 pg mL^{-1} (150 fM) using CNTs labeled with multiple HRPs and secondary antibodies for detection [68]. SWCNT forests

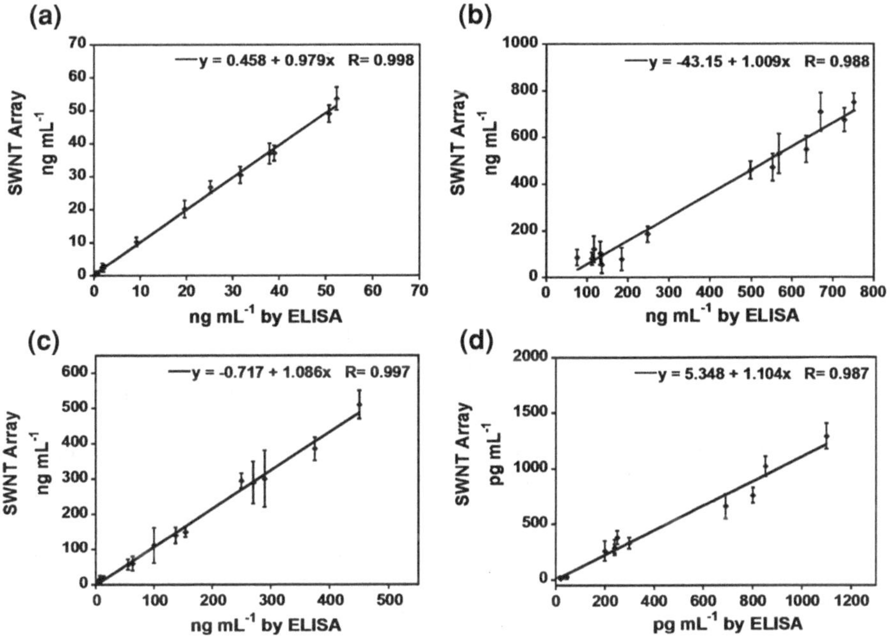

Fig. 1.6 Correlation plots of SWNT immunoarray results for human serum samples against results from ELISA determinations for the same samples (**a**) PSA, (**b**) PSMA, (**c**) PF-4, (**d**) IL-6. Reproduced with permission from [70], copyright American Chemical Society 2006

provided a significant gain in sensitivity over immunosensors without nanotubes because they provide 10–15-fold increase in the number of surface antibodies compared to a flat immunosensor [44]. These sensors gave excellent correlation with ELISA for prostate cancer patient serum using two alternate methods of standardization (Fig. 1.5). These data also demonstrate the efficacy of calf serum as a surrogate for immunosensor standardization in the analysis of human serum samples. The SWCNT sensors were also used to measure attogram PSA levels in cancer cells laser microdissected from prostate tissue. A similar approach was used to obtain a 0.5 pg mL^{-1} DL for IL-6 released from cancer cells into conditioned cell growth media [67].

A 4-electrode SWCNT forest array was used to detect prostate cancer biomarkers PSA, IL-6, platelet factor-4 (PF-4), and prostate specific membrane antigen (PSMA) in the serum of prostate cancer patients and cancer-free controls. High accuracy was confirmed by excellent correlation with results from individual ELISAs giving slopes of correlation plots close to 1.0 and intercepts near zero for all proteins (Fig. 6) [70].

Later, we fabricated AuNP electrodes by depositing a dense layer of 5 nm glutathione-decorated AuNPs onto a 0.5 nm polycation layer onto PG. Excellent sensitivity and DLs were achieved by using 1 μm magnetic bead-Ab_2-HRP bioconjugates with ~7500 HRPs per bead (see Fig. 2 g) [46]. Combining these

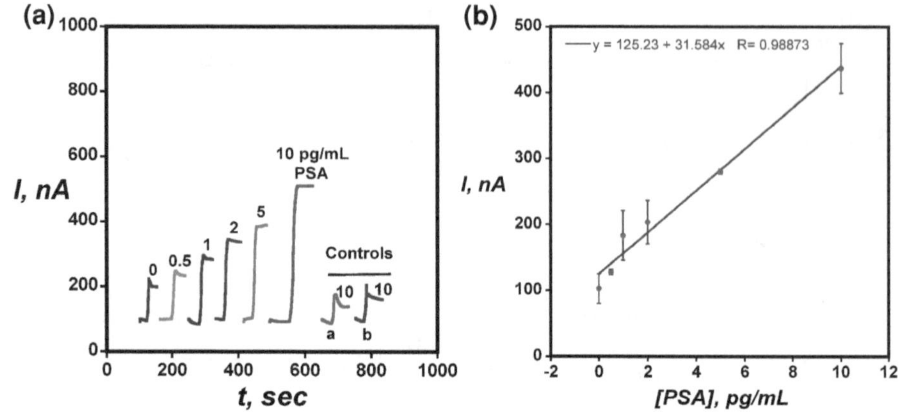

Fig. 1.7 Amperometric responses for AuNP immunosensors at –0.3 V and 3000 rpm in buffer containing 1 mM hydroquinone after injecting 0.04 mM H2O2 to develop the signal (**a**) using Ab2-magnetic bead-HRP with 7500 labels/bead at PSA concentrations shown. Controls: **a** Immunosensors built on bare PG at 10 pg mL-1 PSA (**b**) Immunosensors built on PDDA coated PG surface at 10 pg mL-1 PSA; **b** Influence of PSA concentration on steady state current for AuNP immunosensor using multi-label Ab_2-Magnetic bead-HRP. Reproduced with permission from [46], copyright American Chemical Society 2009

multiply-labeled magnetic beads with the AuNP sensors (Fig. 1.7) gave a DL of 0.5 pg mL^{-1} (20 fM) for PSA. This was eightfold better and sensitivity was fourfold better than SWCNT forest immunosensors. Controls (a) and (b) in Fig. 7 show that AuNPs also provided enhanced sensitivity over flat immunosensors without AuNPs. Both SWCNT forest and AuNP immunosensors gave excellent correlations with ELISA for PSA in cancer patient serum [46, 68].

We also reported a AuNP immunosensor for detection of IL-6 with a DL of 10 pg mL^{-1} in calf serum without using labeled magnetic particles [71]. A comparison under the same assay conditions using human IL-6 cancer biomarker in calf serum revealed that the AuNP immunosensor offers a threefold better detection limit than SWCNT forest immunosensors. In another strategy we used 0.5 μm multi-labeled polymeric beads (polybeads–HRP-Ab_2) to achieve a DL of 10 pg mL^{-1} for MMP-3 [72] in calf serum.

Our most sensitive immunosensor to date is based on the glutathione-protected gold nanoparticle (GSH-AuNP) platform coupled to massively labeled paramagnetic particles (~500,000 HRPs, see Fig. 1.8) for amperometric detection of cancer biomarker interleukin 8 (IL-8). The DL was an unprecedented 1 fg mL^{-1} (100 aM) for IL-8, the lowest protein level yet detected in serum [73]. Accuracy was demonstrated by good correlations with ELISA for determining IL-8 in conditioned growth media from a series of head and neck squamous cell carcinoma (HNSCC) cells. Our detection limit (DL) is similar to that of a DNA barcode method that used PCR amplification before detection to achieve a DL of 1 fg mL^{-1} (30 aM) in goat serum [74].

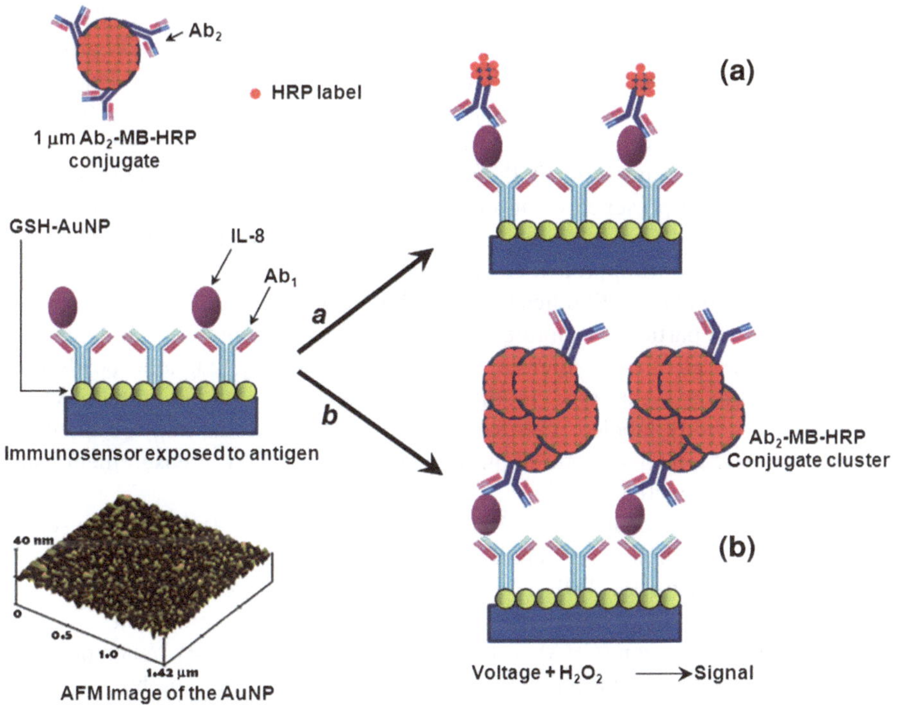

Fig. 1.8 Illustration of detection principles of AuNP immunosensors using a massively labeled strategy. The sensor surface after protein capture is shown on the left at the center. On the bottom left is a tapping mode atomic force microscope image of the AuNP film immunosensor platform. Picture (**a**) on the right shows the immunosensor after treating with biotinylated Ab2 followed by streptavidin modified HRP resulting in HRP-Ab_2 providing 14-16 label per binding event. Picture (**b**) on the right shows the immunosensor after treating with massively labeled Ab_2-MB-HRP particles to obtain amplification by providing ~500,000 enzyme labels per binding event

Other researchers have followed related strategies as described above for detection of IL-6. For example, Wang et. al. [75] reported an amperometric immunosensor to detect interleukin-6 (IL-6) using a AuNP-Poly-dopamine sensor platform and multienzyme-antibody functionalized AuNPs on carbon nanotubes. They obtained a DL of 1 pg mL^{-1} for IL-6 in buffer. Du et. al. [76] used AuNP-modified screen printed carbon electrode to detect p53 phosphorylated at Ser392 (phospho-$p53^{392}$) along with multi-enzyme labeled graphene oxide (GO).

1.3.4 Coupling Nanostructured Surfaces with Electrochemiluminescence (ECL)

Electrochemiluminescence (ECL) is an electrode-driven luminescence process where light emission is initiated by a redox reaction occurring at an electrode, and as such provides luminescence without a light source. Mechanisms, advantages and applications of ECL have been widely reviewed [1, 2]. ECL has grown in importance as a detection method for many types of biomarkers [77–85] and is the basis of several bead-based commercial protein detection instruments [86, 87].

Measurement of proteins using ECL labels is often done using particle-dependent immunoassays [79]. ECL signals proportional to protein concentrations are produced in the presence of an electrolyte solution containing a redox coreactant and measured by a charge-coupled device (CCD) camera or a photomultiplier tube (PMT). This approach can be used in various types of sandwich assays. For example, secondary antibodies linked with ECL labels [e.g., $Ru(bpy)_3^{2+}$] can be immobilized on a particle to capture the analyte protein, then collected by capture antibodies on an electrode. Alternatively, capture-antibody-magnetic beads with streptavidin attached can bind to the protein, and then recruit a biotinylated monoclonal antibody labeled with $Ru(bpy)_3^{2+}$. After NSB blocking the magnetic particles are magnetically captured onto an electrode for ECL measurement using a suitable co-reactant [88].

Selected small molecules, ions [89–94] or enzymes [95–98] can be used as coreactants. For example, acetylcholinesterase was utilized as coreactant to detect tumor necrosis factor-α (TNF-α) on a gold electrode to achieve a DL of $\sim$3 pg mL^{-1} [96]. In another study $S_2O_8^{2-}$ was used as coreactant to detect carcinoembryonic antigen (CEA) with a DL of 0.03 pg mL^{-1} [94]. When potential was scanned in a negative direction, CdSe–CdS nanoparticles immobilized on the electrode were reduced to CdSe–CdS–$^{\bullet}$. The reduced form of $S_2O_8^{2-}$ ($SO_4^{\bullet}$) further reacted with the CdSe–CdS–$^{\bullet}$ to provide excited state (CdSe–CdS*) that generated ECL. Detection of CEA was based on steric hindrance due to formation of the immunocomplex, which inhibited the transfer of electrons and $S_2O_8^{2-}$ to the electrode surface leading to a decrease in ECL intensity.

Tripropylamine (TPrA) is a commonly used coreactant for $Ru(bpy)_3^{2+}$ labels since the $Ru(bpy)_3^{2+}$/TPrA ECL system provides high sensitivity [79, 82]. This system has been used to detect cancer biomarkers such as PSA, cancer antigen 125 (CA-125, ovarian cancer), P53 protein, and others [88–91, 93]. ECL emission from the $Ru(bpy)_3^{2+}$/TPrA system as a function of applied potential consists of two complex redox pathways that provide ECL emission from the excited state $Ru(bpy)_3^{2+*}$ [77, 79, 82].

A particularly useful ECL pathway for detecting low concentrations of proteins by immunosensors is initiated by oxidation at 0.9 V vs SCE of the sacrificial reductant TprA, whose products react in a complex pathway with $Ru(bpy)_3^{2+}$ to yield $Ru(bpy)_3^{2+*}$. We developed an immunosensor on a SWCNT forest platform for PSA in serum utilizing this approach with $Ru(bpy)_3^{2+}$-silica nanoparticles attached to secondary antibodies (RuBPY-silica-Ab_2) as labels [99]. Addition of surfactants increases the hydrophobicity of the sensor surface via an adsorbed surfactant layer, which facilitates oxidation of TprA [100]. Including Triton X-100 and Tween 20 in the electrolyte solution containing TPrA improved PSA sensitivity tenfold compared to TprA in surfactant-free solutions [99]. Surface

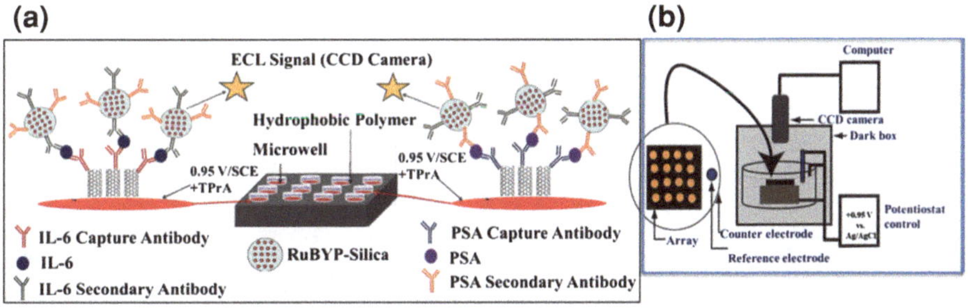

Fig. 1.9 System for ECL immunoarray: **a** discrete wells in red on a 1 x 1 in. pyrolytic graphite chip (*on left, black*). SWCNT forests are surrounded by hydrophobic polymer (*white*) to make microwells on the chip. Wells have SWCNT forests in their bottoms decorated with primary antibodies. Wells are filled with sample solutions and incubated to capture the analyte proteins. After washing, RuBPY-silica nanoparticles with cognate secondary antibodies are added and bind to the captured protein analytes. Appropriate washing with blocking buffers minimizes non-specific binding. **b** The chip is placed in an open top electrochemical cell, 0.95 V vs. SCE is applied, and ECL is measured with a CCD camera

hydrophobicity prevents deprotonation of TPrA cationic radical formed by oxidation of TprA to promote increased ECL.

We utilized these considerations to design an ECL immunosensor array on a 1×1 in. pyrolytic graphite chip (Fig. 1.9) [101]. The array featured SWCNT forests self-assembled in the bottoms of 10 μL wells made by painted-on polymer walls, which can be visualized by AFM (Fig. 1.10). RuBPY-silica-Ab_2 (100 nm dia.) nanoparticles with antibodies to both PSA and IL-6 attached were used for detection. Antibodies on SWCNTs in the wells capture analyte proteins from 5 μL of sample. The RuBPY-silica-Ab_2 particles are then added to bind to the proteins captured on the arrays. After appropriate NSB blocking, detection is initiated by electrochemical oxidation of tripropylamine (TprA), which generates emission of ECL from $[Ru(bpy)_3]^{2+}$ in the nanoparticles. ECL is measured with a CCD camera with array chip in an open top electrochemical cell in a dark box (Fig. 1.9). The hydrophobic polymer walls confine liquids in the analytical wells to enable simultaneous assays of different proteins in serum while avoiding cross-contamination (Fig. 1.11a–d). DLs in serum were 1 pg mL^{-1} for PSA and 0.25 pg mL^{-1} for IL-6 [101]. Array measurements of these biomarker proteins in prostate cancer patient serum gave good correlations with single-protein ELISAs (Fig. 1.11e, f).

Amplification with other nanoparticles as ECL labels has also been demonstrated [102–108]. CdSe nanocrystals (NCs) and Qdots have been used for immunoassays. For example, an ECL immunosensor was constructed by immobilizing CdS Qdots and capture antibodies on a poly(diallyldimethylammonium chloride)-functionalized CNT-modified (PDDA/CNTs) electrode to detect α-fetoprotein [98] A bio-bar-code probe labeled antibody was designed by conjugation of hemin and a single-stranded guanine-rich oligonucleotide to the antibody on gold nanoparticles. The bio-bar-coded probe was captured on the

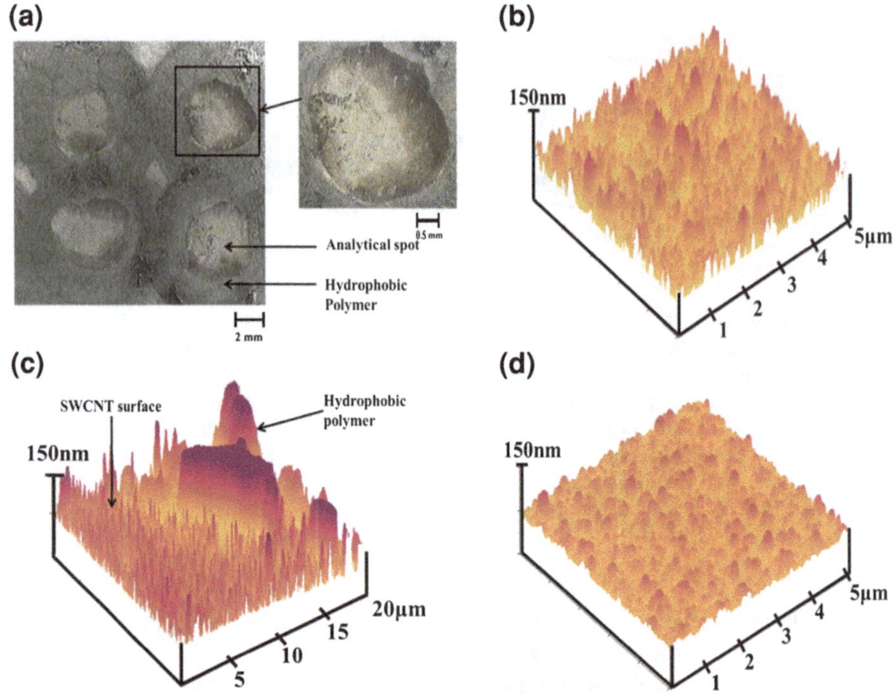

Fig. 1.10 Microscopy of microwell arrays: **a** Optical micrograph of 4 spots on a pyrolytic graphite array showing the light green hydrophobic polymer wall surrounding SWCNT forest spots. The inset shows a single SWCNT well surrounded by hydrophobic polymer. (**b** to **d**) are tapping mode atomic force microscope images of films on mica: **b** dense SWCNT forest in the bottom of an analytical well; **c** view showing the polymer wall and the adjacent SWCNT forest; **d** SWCNT forest in the bottom of a microwell after covalent linkage of 2 nmol mL^{-1} anti-PSA antibody in pH 7.0 PBS buffer + 0.05% Tween-20 followed by washing with PBS buffer. Reproduced with permission from ref. [101], copyright American Chemical Society 2011

immunosensor surface. The α-fetoprotein was detected with a linear range of 0.01 pg mL^{-1} to 1 ng mL^{-1}. In another study an ECL-based immunoassay was developed combining CdTe Qdots with ultra-thin nanoporous gold leaf electrodes [109] using $S_2O_8^{2-}$ as coreactant, to provide a DL of 0.01 mg mL^{-1} for CEA [92].

1.4 Nanostructured Protein Sensors in Microfluidic Arrays

Microfluidics coupled to bioanalytical devices has the potential to improve multiplexing and signal/noise, consume less expensive reagents and provide a degree of automation. In this section, we briefly summarize recent efforts to couple microfluidics to nanoparticle-based protein immunoassays for multiplexed biomarker detection. A recent example involves a 16-sensor electrochemical chip

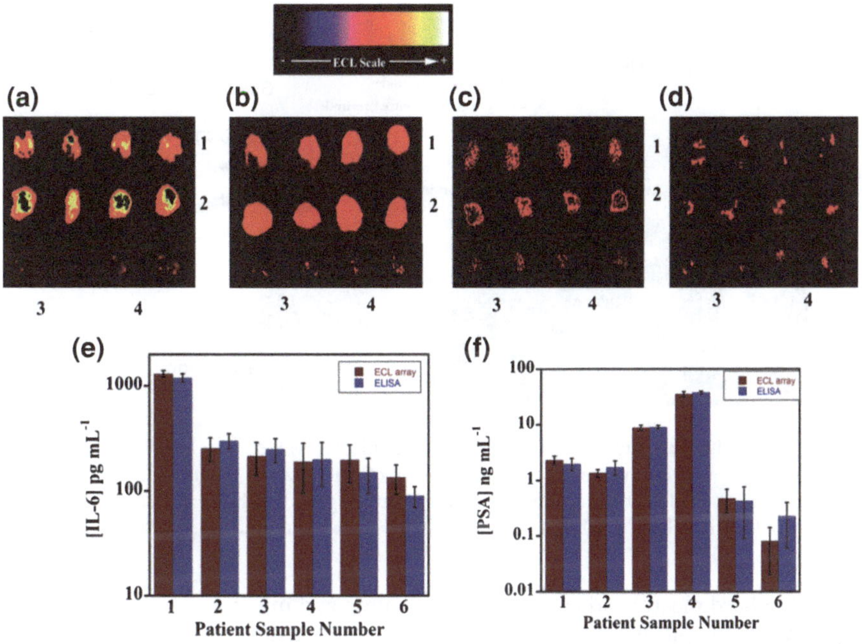

Fig. 1.11 Microwell array ECL images showing detection of PSA and IL-6 in mixtures in calf serum, obtained at 0.95 V vs Ag/AgCl using 0.05% Tween 20 + 0.05% Triton-X 100 + 100 mM TprA, pH 7.5. RuBPY-silica nanoparticles were used with antibodies attached for both proteins: **a** (1) 5 ng mL-1 PSA, (2) 1 ng mL-1 IL-6:, **b** (1) 0.4 ng mL-1 PSA, (2) 0.2 ng mL-1 IL-6, (**c**) (1) 40 pg mL-1 PSA, (2) 20 pg mL-1 IL-6, and (**d**) (1) 1 pg mL-1 PSA, 2) 0.25 pg mL-1 IL-6. In all images, controls are indicated by duplicate spots (3) 0 pg mL-1 IL-6, and (4) 0 pg mL-1 PSA. E and F are comparison of ECL array determinations of PSA and IL-6 in patient serum with individual ELISAs: **e** IL-6; **f** PSA. Samples 1 to 4 from prostate cancer patients; samples 5 and 6 were from cancer-free patients. Reproduced with permission from ref. [101] copyright American Chemical Society 2011

using sensors coated with a DNA dendrimer/conducting polymer film decorated with capture antibodies that gave high sensitivity [110]. Oral cancer protein markers IL-8 and IL-1b as well as the RNA biomarker IL-8mRNA were measured using HRP-labeled secondary antibodies for detection. DLs in buffer of 100–200 fg mL^{-1} were obtained for the proteins and a DL of 10 aM was achieved for IL-8 mRNA. Poorer DLs were found in human saliva, i.e. 4 fM IL-8mRNA and 7.4 pg mL^{-1} IL-8 [111]. Statistical performance evaluation using assays data from saliva of oral cancer patients predicted 90 % clinical sensitivity and specificity for tests involving IL-8 mRNA and IL-8.

We recently coupled nanoparticle-based sensors on an 8-biosensor array with off-line protein capture into a simple microfluidic system (Fig. 1.12) [112]. This microfluidic immunoassay system features AuNP sensor electrodes built on a screen-printed carbon platform inserted into a molded 70 μL PDMS channel

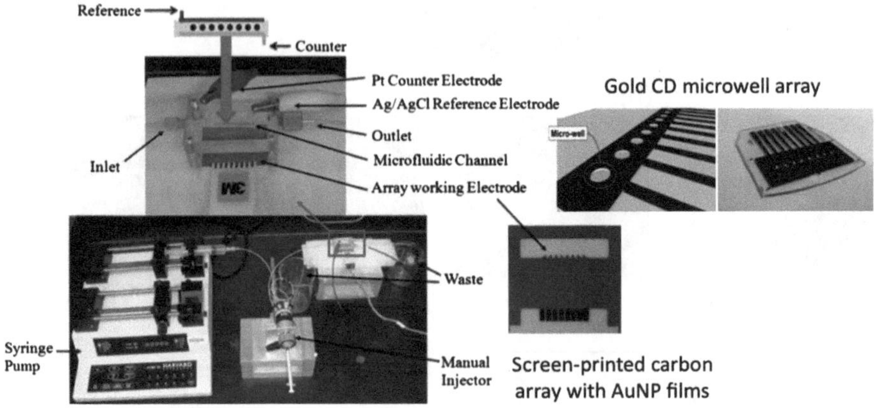

Fig. 1.12 Microfluidic system consisting of pump, injector valve, and insertable 8-electrode arrays in a 70 μL PDMS channel. **a** and **b** are views of a gold array featuring 1 μL microwells fabricated from a gold CD by computer template printing and etching. **c** is a screen-printed carbon array (Kanichi Ltd., UK) that has been coated with 5 nm gold nanoparticles

enclosed in hard plastic and equipped with a pump and injector valve. We used this system with off-line protein capture by magnetic particles linked to secondary antibodies and 200,000 HRP labels to achieve sub pg mL^{-1} DLs for proteins in serum. Additional features include multiplexing, speed (~ 1 h/assay), and low cost.

Figure 1.13 shows calibration data obtained for the simultaneous detection of PSA and IL-6 in diluted serum using the microfluidic system in Fig. 1.12. The assay begins with off-line capture of the proteins by the labeled magnetic particles, after which the particles are magnetically separated and washed. The injector sample valve is used to inject these particles into the detection chamber, and flow is stopped for 15 min. The particles that have captured analyte proteins bind to the capture antibodies on the sensors. Flow is resumed, NSB is minimized by blocking agents, and a mixture of H_2O_2 and hydroquinone is injected to develop the signal (Scheme 1.1). The device gives peaks with excellent signal/noise in the sub-pg mL^{-1} range [112] (Fig. 1.13). Excellent dynamic ranges and DLs of ~0.2 pg mL^{-1} were obtained for both proteins in mixtures. The screen-printed electrode chips are used once, then discarded, and a new chip is inserted into the device for the subsequent assay.

While the screen-printed sensor chips we use are inexpensive (Kanichi, UK, ~ $10 ea.), we are also exploring alternative methodologies to make chips that can be interfaced with the microfluidic system in Fig. 1.12. For example, ink-jet printing was used to print 8-electrode arrays from gold nanoparticle ink onto Kapton plastic at a cost of about $0.2/chip [113]. We also made gold arrays from gold compact discs (CDs) featuring microwells around the sensor electrodes (Fig. 1.12, on right) [114]. The gold CD sensor arrays were fabricated at a similar cost in materials by thermal transfer of laser jet toner from computer-printed patterns and selective chemical etching. The resulting sensor surfaces retain the nm-sized CD

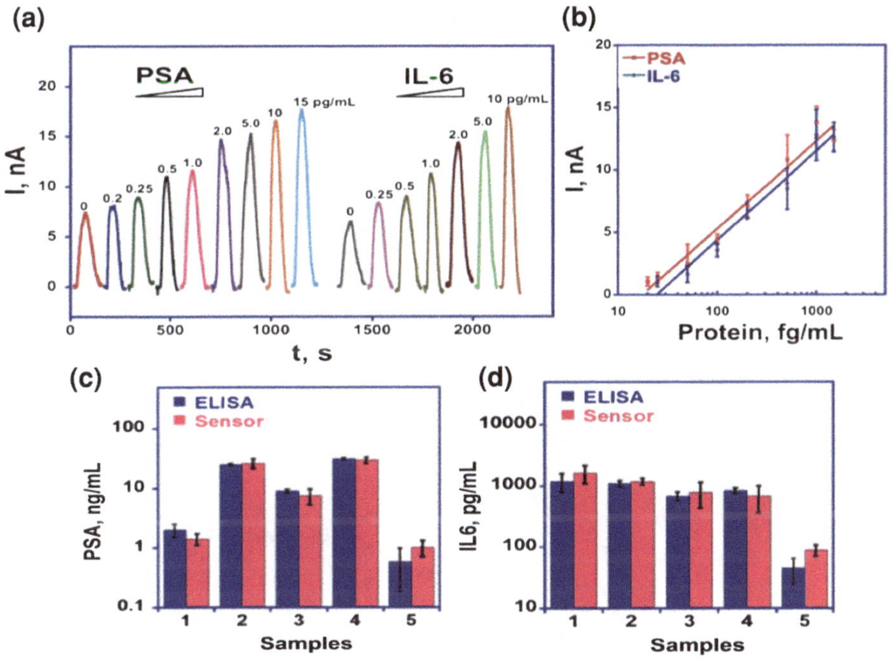

Fig. 1.13 Calibration of 8-sensor microfluidic array with off-line analyte capture by multilabel HRP-MP-Ab2 particles using 200,000 HRP labels/particle for PSA and IL-6 mixtures in serum. Signals developed by injecting 1 mM hydroquinone mediator + 100 μM hydrogen peroxide as enzyme activator. **c** Simultaneous determinations by the array compared to individual ELISAs for PSA and IL-6 in patient serum: 1-4 are from prostate cancer patients; 5 is a cancer-free control. Reproduced with permission from ref. 112 copyright Elsevier, 2011

grooves (Fig. 1.14). These arrays were integrated into the microfluidic device for electrochemical detection of interleukin-6 (IL-6) in diluted serum. Capture antibodies were attached onto the sensors, and a biotinylated detection antibody attached to polymerized HRP (polyHRP) was used for signal amplification. DL for IL-6 in diluted serum was 10 fg mL^{-1} (385 aM). These easily fabricated, ultrasensitive immunoarrays have some advantages over our previous screen-printed varieties. They achieved excellent sensitivity without inclusion of gold nanoparticle films or use of off-line protein capture. This decreases the length of the assay protocol, and also avoids synthesis and characterization of MP bioconjugates.

1.5 Conclusions and Future Perspectives

This chapter summarizes recent progress in development of ultrasensitive electrochemical devices to measure cancer biomarker proteins. The emphasis is on the use of nanoparticles and nanostructured sensors aimed for use in clinical cancer

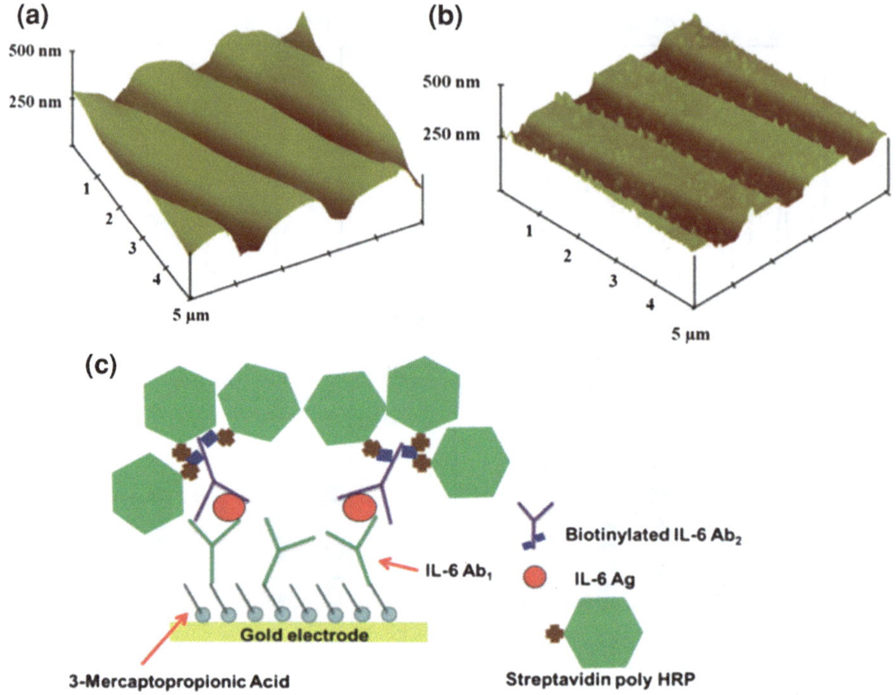

Fig. 1.14 Arrays made from gold CDs. Tapping mode AFM images of (**a**) exposed bare gold CD-R surface (**b**) Anti IL-6 capture antibody attached to the gold CD-R surface. **c** Amplification strategy using streptavidin poly-HRP. The streptavidin poly-HRP attaches to biotinylated anti-human IL-6 detection antibody bound to IL-6 on the sensor before the measurement step. Reproduced with permission from Ref. [114] copyright Royal Society of Chemistry, 2011

diagnostics, and focuses largely on work from our own laboratory. Low cost, reliable multiplexed protein detection devices have great promise for future cancer diagnostics since current clinical practice often involves single biomarkers with low predictive power, qualitative physical examinations, or biopsies coupled with pathological identification of the cancer. Certainly, measurement of a reliable panel of multiple biomarker proteins in blood or saliva would be an enormous advance for cancer diagnostics, individualized therapy, and lowering of patient stress.

As we have tried to point in this chapter, development of an ultrasensitive measurement device is only the first step in realizing broad based use of multiple-protein detection in cancer diagnostics. As illustrated in Figs. 1.5, 1.6, 1.11 and 1.13, an immediate concern is accuracy in the sample medium to be used in the diagnostic test. That is, the new device should be tested with real samples against accurate alternative methods such as ELISA, and good correlations obtained to ensure accuracy.

The next important issue is the choice of the biomarker protein panel. Reliability of the panels for diagnostics needs to be quantitatively examined by studies on large numbers of patient samples to develop panels with a high degree of reliability. Progress in this area has been relatively slow. However, in the few studies that have been completed, reliable panels with higher clinical sensitivity and selectivity seem to be emerging.

In our research, we have demonstrated the power of combining nanostructured electrodes with multilabel particles or polymers to provide ultrahigh sensitivity, accuracy, flexibility, and low cost. The advantages of combining multiplexed immunoassays with microfluidics are realized in semi-automation, lower cost and reagent consumption, improved speed, and better signal/noise. Further automation and simplification of microfluidic systems is needed, however, for protein measurement devices to reach POC applications in the clinic.

Finally, we frequently hear criticism of biomarker discovery research for its slowness in yielding benefits to the clinic [21, 22]. This lag is due partly to the absence of reliable low cost protein detection devices, partly to stubborn reliance by some on single new biomarkers when multiple biomarker panels are clearly needed, and partly to the development time required for basic research to be technically adapted for the clinic. A case for comparison is the electrochemical blood glucose sensor, now the method of choice for diabetic patient home use. In this case, it was already well known that glucose was an important biomarker for diabetes. Yet it took roughly a decade from the first paper on the mediated electrochemical glucose biosensor [115] to finally make it to market. For cancer biomarker proteins, we have a number of very low concentration analytes to measure for each cancer, and we really don't know exactly which ones are the best yet. In addition, devices designed to diagnose the most common cancers may be required to measure up to 100 or more biomarkers. Clearly, there's a lot of research and development yet to be done, and its unrealistic to expect that the entire task can be completed within the next few years. However, simpler devices designed to diagnose single cancers could appear in shorter periods. As this chapter documents, significant progress is being made and the payoff in advanced diagnostic ability is enormous.

Acknowledgments This work was supported by NIH grants ES013557 from NIEHS and EB014586 from NIBIB (JFR), by a Walton Research Fellowship to JFR from Science Foundation Ireland, and by grant P20RR016457 from NCRR/NIH (BSM). The authors thank collaborators and research students named in joint publications for their excellent contributions to the project, without which progress would not have been possible.

References

1. Manne, U., Srivastava, R.G., Srivastava, S.: Recent advances in biomarkers for cancer diagnosis and treatment. Drug Discov. Today **10**, 965–976 (2005)
2. Rusling, J.F., Kumar, C.V., Gutkind, J.S., et al.: Measurement of biomarker proteins for point-of-care early detection and monitoring of cancer. Analyst **135**, 2496–2511 (2010)
3. Ludwig, J.A., Weinstein, J.N.: Biomarkers in cancer staging, prognosis and treatment selection. Nat. Rev. Cancer **5**, 845–856 (2005)

4. Kulasingam, V., Diamandis, E.P.: Strategies for discovering novel cancer biomarkers through utilization of emerging technologies. Nat. Clin. Pract. Oncol. **5**, 588–599 (2008)
5. Hanash, S.M., Pitteri, S.J., Faca, V.M.: Mining the plasma proteome for cancer biomarkers. Nature **452**, 571–579 (2008)
6. Giljohan, D.A., Mirkin, C.A.: Drivers of biodiagnostic development. Nature **462**, 461–464 (2009)
7. Hanash, S.M., Baik, C.S., Kallioniemi, O.: Emerging molecular biomarkers—blood-based strategies to detect and monitor cancer. Nat. Rev. Clin. Oncol. **8**, 142–150 (2011)
8. Wang, J.: Nanomaterial-based electrochemical biosensors. Analyst **130**, 421–426 (2005)
9. Wang, J., Katz, E., Willner, I.: Biomaterial-nanoparticle hybrid systems for sensing and electronic devices. In: Katz, E., Willner, I. (eds.) Bioelectronics: from Theory to Applications, pp. 231–264. Wiley, Weinheim (2005)
10. Kim, S.N., Rusling, J.F., Papadimitrakopolous, F.:Carbon nanotubes in electronic and electrochemical detection of biomolecules. Adv. Mater. **19**, 3214–3228 (2007)
11. Patolsky, F., Zheng, G., Lieber, C.M.: Nanowire-based biosensors. Anal. Chem. **78**, 4260–4269 (2006)
12. Atkinson, A.J., Colburn, W.A., DeGruttola, V.G., et al.: Biomarkers and surrogate endpoints: preferred definitions and conceptual framework. Clin. Pharmacol. Ther. **69**, 89–95 (2001)
13. Kulasingam, V., Diamandis, E.P.: Strategies for discovering novel cancer biomarkers through utilization of emerging technologies. Nat. Clin. Prac. Oncol. **5**, 588–599 (2008)
14. Hawkridge, A.M., Muddiman, D.C.: mass spectrometry–based biomarker discovery: toward a global proteome index of individuality. Ann. Rev. Anal. Chem. **2**, 265–277 (2009)
15. Wang, J.: Electrochemical biosensors: towards point-of-care cancer diagnostics. Biosens. Bioelectron. **21**, 1887–1892 (2006)
16. Wagner, P.D.,Verma, M., Srivastava, S.: Challenges for biomarkers in cancer detection. Ann. N. Y. Acad. Sci. **1022**, 9–16 (2004)
17. Li, J., Zhang, Z., Rosenzweig, J., et al.: Proteomics and bioinformatics approaches for identification of serum biomarkers to detect breast cancer. Clin. Chem. **48**, 1296–1304 (2002)
18. Ward, M.A., Catto, J.W.F., Hamdy, F.C.: Prostate specific antigen: biology, biochemistry and available commercial assays. Ann. Clin. Biochem. **38**, 633–651 (2001)
19. Tothill, I.E.: Biosensors for cancer markers diagnosis. Semin. Cell Dev. Biol. **20**, 55–62 (2009)
20. Choi, Y.-E., Kwak, J.-W., Park, J.W.: Nanotechnology for early cancer detection. Sensors **10**, 428–455 (2010)
21. Mischak, H., Allmaier, G., Apweiler, R., et al.: Recommendations for biomarker identification and qualification in clinical proteomics. Sci. Transl. Med. **2**, 46ps42 (2010)
22. Goodsaid, F.M., Mendrick, D.L.: Translational medicine and the value of biomarker qualification. Sci. Transl. Med **2**, 47ps44 (2010)
23. Kingsmore, S.F.: Multiplexed protein measurement: technologies and applications of protein and antibody arrays. Nat. Rev. Drug Discov. **5**, 310–320 (2006)
24. Rasooly, A., Jacobson, J.: Development of biosensors for cancer clinical testing. Biosens. Bioelectron. **21**, 1851–1858 (2006)
25. Heineman, W.R., Halsall, H.B.: Strategies for electrochemical immunoassay. Anal. Chem. **75**, 1321A–1331A (1985)
26. Vijayawardhana, C.A., Halsall, H.B., Heineman, W.R.: Milestones of electrochemical immunoassay at Cincinnati. In: Chambers, J.Q., Bratjer-Toth, A. (eds.) Electroanalytical Methods for Biological Materials, pp. 195–231. Marcel Dekker, NY (2002)
27. Ronkainen-Matsuno, N.J., Thomas, J.H. Halsall H.B., Heineman, W.R.: Electrochemical immunoassay moving into the fast lane. Trends Anal. Chem. **21**, 213–225 (2002)
28. Bange, A. Halsall, H.B., Heineman, W.R.: Microfluidic immunosensor systems. Biosens. Bioelectron. **20**, 2488–2503 (2005)

29. Ronkainen-Matsuno, N.J., Halsall H.B., Heineman, W.R. Electrochemical biosensors. Chem. Soc. Rev. **39**, 1747–1763 (2010)
30. Warsinke, A., Stocklein, W., Leupold, E., Micheel, E., Scheller, F.W.: Electrochemical immunosensors on the road to proteomic chips. In: Perspectives in Bioanalysis, vol. 1. Elsevier, Amsterdam (2007)
31. Lu, B., Smyth, M.R., O'Kennedy, R.: Immunological activities of IgG antibody on pre-coated Fc receptor surfaces. Anal. Chim. Acta **331**, 97–102 (1996)
32. Carter, R.M.M.A. Poli, M. Pesavento, D.E.T. Sibley, G.J. Lubrano, G.G.: Guilbault, immunoelectrochemical biosensors for detection of saxitoxin and brevetoxin. Immunomethods **3**, 128–133 (1993)
33. Warsinke, A., Benkert, A., Scheller, F.W. Electrochemical immunoassays. Fresenius J. Anal. Chem. **366**, 622–634 (2000)
34. Yakovleva, J., Emneus, J.: Electrochemical immunoassays. In: Bartlett, P.N. (ed.) Handbook of Bioelectrochemistry, pp. 377–410. Wiley, New York (2008)
35. Wilson, M.S.: Electrochemical immunosensors for the simultaneous detection of two tumor markers. Anal. Chem. **77**, 1496–1502 (2005)
36. Wilson, M.S., Nie, W.: Electrochemical multianalyte immunoassays using an array-based sensor. Anal. Chem. **78**, 2507–2513 (2006)
37. Wilson, M.S., Nie, W.: Multiplex measurement of seven tumor markers using an electrochemical protein chip. Anal. Chem. **78**, 6476–6483 (2006)
38. Wang, J.: Nanoparticle-based electrochemical bioassays of proteins. Electroanalysis **19**, 769–776 (2007)
39. Veetil, J.V., Ye, K (2007) Development of immunosensors using carbon nanotubes. Biotechnol. Prog. **23**, 517–531 (2007)
40. Luo, X., Morrin, A., Killard, A.J., Smyth, M.R.: Application of nanoparticles in electrochemical sensors and biosensors. Electroanalysis **18**, 319–326 (2006)
41. Zhang, H., Zhao, Q., Li, X.-F., Le, X.C.: Ultrasensitive assays for proteins. Analyst **132**, 724–737 (2007)
42. Wang, J.: Nanomaterial-based amplified transduction of biomolecular interactions. Small **1**, 1036–1043 (2005)
43. Rusling, J.F., Yu, X., Munge, B.S., Kim, S.N., Papadimitrakopoulos, F. In: Davis, J. (ed.) Engineering the Bioelectronic Interface, pp. 94–118. Royal Society of Chemistry, UK (2009)
44. Malhotra, R., Papadimitrakopoulos, F., Rusling. J. F.: Sequential layer analysis of protein immunosensors based on single wall carbon nanotube forests. Langmuir **26**, 15050–15056 (2010)
45. Das, J., Kelley, S.O.: Protein detection using arrayed microsensor chips: tuning sensor footprint to achieve ultrasensitive readout of CA-125 in serum and whole blood. Anal. Chem. **83**, 1167–1172 (2011)
46. Mani, V., Chikkaveeraiah, B.V., Patel, V., Gutkind, J.S., Rusling, J.F.: Ultrasensitive immunosensor for cancer biomarker proteins using gold nanoparticle film electrodes and multienzyme-particle amplification. ACSNano **3**, 585–594 (2009)
47. Soleymani, L., Fang, Z., Sargent, E.H., Kelley, S.O.: Programming the detection limits of biosensors through controlled nanostructuring. Nat. Nanotech. **4**, 844–848 (2009)
48. Cai, D., Ren L., Zhao H., Xu C., Zhang L., et. al.: A molecular-imprint nanosensor for ultrasensitive detection of proteins. Nat. Nanotechn. **5**, 597–601 (2010)
49. Osakai, T., Yuguchi, Y., Gohara, E., Katano, H.: Direct label-free electrochemical detection of proteins using the polarized oil/water interface. Langmuir **26**, 11530–11537 (2010)
50. Genc, R., Murphy, D., Fragoso, A., Ortiz, M., O'Sullivan, C. K.: Signal-enhancing thermosensitive liposomes for highly sensitive immunosensor development. Anal. Chem. **83**, 563–570 (2011)
51. Zhao, Y.L., Stoddart, J.F.: Noncovalent functionalization of single-walled carbon nanotubes. Acc. Chem. Res. **42**, 1161–1171 (2009)
52. Jacobs, C.B., Peairs, M.J., Venton, B.J.: Carbon nanotube based electrochemical sensors for biomolecules. Anal. Chim. Acta. **662**, 105–127 (2010)

53. Dequaire, M., Degrand, C., Limoges, B.: An electrochemical metalloimmunoassay based on a colloidal gold label. Anal. Chem. **72**, 5521 (2000)
54. Guo, H., He, N., Ge, S., Yang, D., Zhang, J.: MCM-41 mesoporous material modified carbon paste electrode for the determination of cardiac troponin I by anodic stripping voltammetry. Talanta **68**, 61–66 (2005)
55. Velev, O.D., Kaler, E.W.: In situ assembly of colloidal particles into miniaturized biosensors. Langmuir **15**, 3693–3698 (1999)
56. Liu, G., Wang, J., Kim, J., Jan, M., Collins, G.: Electrochemical coding for multiplexed immunoassays of proteins. Anal. Chem. **76**, 7126–7130 (2004)
57. Daniels, J.S., Pourmanda, N.: label-free impedance biosensors: opportunities and challenges. Electroanalysis **19,** 1239–1257 (2007)
58. Tkac, J., Davis, J.J.: Label-free field effect protein sensing. In: Davis, J.J. (ed.) Engineering the Bioelectronic Interface, pp. 193–224. Royal Society of Chemistry, UK (2009)
59. Berggren, C., Bjarnason, B., Johansson, G. An immunological Interleukine-6 capacitive biosensor using perturbation with a potentiostatic step. Biosens. Bioelectron. **13**, 1061–1068 (1998)
60. Bart, M., Stigter, E.C.A., Stapert, H.R., de Jong, G.J., van Bennekom, W.P. On the response of a label-free interferon-γ immunosensor utilizing electrochemical impedance spectroscopy, Biosens. Bioelectron. **21**, 49–59 (2005)
61. Wang, J., Liu, G., Jan, M.R.: Ultrasensitive electrical biosensing of proteins and DNA: carbon-nanotube derived amplification of the recognition and transduction events. J. Am. Chem. Soc. **126**, 3010–3011 (2004)
62. Munge, B., Liu, G., Collins, G., Wang, J.: Multiple enzyme layers on carbon nanotubes for electrochemical detection down to 80 DNA copies. Anal. Chem. **77**, 4662–4666 (2005)
63. Wang, J., Liu, G., Munge, B., Lin, L., Zhu, Q. Angew. Chem. Int. Ed. **43**, 2158–2161 (2004)
64. Rusling, J. F., Sotzing, G., Papadimitrakopoulos, F.: Designing nanomaterials-enhanced electrochemical immunosensors for cancer biomarker proteins, Bioelectrochem. **76**, 189–194 (2009)
65. Yu, X., Munge, B. Patel, V., Jensen, G., Bhirde, A., Gong, J.D., Kim, S.N., Gillespie, J. Gutkind, J.S., Papadimitrakopoulos, F., Rusling, J.F.: Carbon nanotube amplification strategies for highly sensitive immunosensing of cancer biomarkers in serum and tissue. J. Am. Chem. Soc. **128**, 11199–11205 (2006)
66. Munge, B.S., Krause, C.E., Malhotra, R., Patel, V., Gutkind, J.S., Rusling, J.F. Electrochemical immunosensors for Interleukin-6. Comparison of carbon nanotube forest and gold nanoparticle platforms. Electrochem. Comm. **11**, 1009–1012 (2009)
67. Malhotra, R., Patel, V., Vaqué, J.P., Gutkind, J.S., Rusling, J.F.: Ultrasensitive electrochemical immunosensor for oral cancer biomarker IL-6 using carbon nanotube forest electrodes and multilabel amplification. Anal. Chem. **82**, 3118–3123 (2010)
68. Yu, X., Kim, S.N., Papadimitrakopoulos, F., Rusling, J.F., Protein immunosensor using single-wall carbon nanotube forests with electrochemical detection of enzyme labels. Molec. Biosys. **1**, 70–78 (2005)
69. Jensen, G.C., Yu, X., Munge, B., Bhirde, A., Gong, J.D., Kim, S.N., Papadimitrakopoulos, F. Rusling, J.F.: Characterization of multienzyme-antibody-carbon nanotube bioconjugates for immunosensors. J. Nanosci. Nanotechnol. **9**, 249–255 (2009)
70. Chikkaveeraiah, B.V. Bhirde, A., Malhotra, R., Patel, V., Gutkind, J.S., Rusling, J.F.: Single-wall carbon nanotube forest immunoarrays for electrochemical measurement of 4 protein biomarkers for prostate cancer. Anal. Chem. **81**, 9129–9134 (2009)
71. Munge, B.S., Krause, C.E., Malhotra, R., Patel, V., Gutkind, J.S., Rusling, J.F.: Electrochemical immunosensors for Interleukin-6. Comparison of carbon nanotube forest and gold nanoparticle platforms. Electrochem. Comm. **11**, 1009–1012 (2009)
72. Munge, B.S., Fisher, J., Millord, L.N., Krause, C.E., Dowd, R.S., Rusling, J.F.: Sensitive electrochemical immunosensor for matrix metalloproteinase-3 based on single-wall carbon nanotubes. Analyst **135**, 1345–1350 (2010)

73. Munge, B.S., Coffey, A.L., Doucette, J.M., Somba, B.K., Malhotra, R., Patel, V., Gutkind, J.S., Rusling, J.F.: Nanostructured immunosensor for attomolar detection of cancer biomarker Inerleukin-8 using massively labelled superparamagnetic particles. Angew. Chem. Int. Ed. **50**, 7915–7918 (2011)
74. Nam, J.-M., Thaxton, C.S., Mirkin, C.A.: Nanoparticle-based bio-bar codes for ultrasensitive detection of proteins. Science **301**, 1884–1886 (2003)
75. Wang, G., Huang, H., Zhang, G., Zhang, X., Fang, B., Wang, L. Dual amplificatoin strategy for the fabrication of highly sensitive interleukin-6 amperometric immunosensor based on poly-dopamine. Langmuir **27**, 1224–1231 (2011)
76. Du, D., Wang, L., Shao, Y., Wang, J., Engelhard, M.H., Lin, Y.: Functionalized graphene oxide as a nanocarrier in a multienzyme labeling amplification strategy for ultrasensitive electrochemical immunoassay of phosphorylated p53 (S392). Anal. Chem. **83**, 746–752 (2011)
77. Bard, A.J. (ed.): Electrogenerated Chemiluminescence. Marcel Dekker, New York (2004)
78. Gorman, B.A., Francis, P.S., Barnett, N.W.: Tris(2,29-bipyridyl)ruthenium(II) chemiluminescence. Analyst **131**, 616–639 (2006)
79. Miao, W.J.: Electrogenerated chemiluminescence and its biorelated applications. Chem. Rev. **108**, 2506–2553 (2008)
80. Marquette, C.A., Blum, L.J.: Electro-chemiluminescent biosensing. Anal. Bioanal. Chem. **390**, 155–168 (2008)
81. Bertoncello, P., Forster, R.J.: Nanostructured materials for electrochemiluminescence (ECL)-based detection methods: recent advances and future perspectives. Biosens. Bioelectron. **24**, 3191–3200 (2009)
82. Forster, R.J., Bertoncello, P., Keyes, T.E.: Electrogenerated chemiluminescence. Annu. Rev. Anal. Chem. **2**, 359–385 (2009)
83. Qi, H., Peng, Y., Gao, Q., Zhang, C.: Applications of nanomaterials in electrogenerated chemiluminescence biosensors. Sensors **9**, 674–695 (2009)
84. Hu, L., Xu, G.: Applications and trends in electrochemiluminescence. Chem. Soc. Rev. **39**, 3275–3304 (2010)
85. Wei, H. Wang, E.: Electrochemiluminescence of tris(2,2′-bipyridyl)ruthenium and its applications in bioanalysis: a review. Luminescence **26**, 77–85 (2011)
86. Roche Diagnostics: http://rochediagnostics.ca/lab/solutions/e2010.php
87. Meso Scale Diagnostics: www.mesoscale.com
88. Debad, J.B., Glezer, E.N., Leland, J.K., Sigal, G.B., Wholstadter, J. In: Bard, A.J. (ed.) Electrogenerated Chemiluminescence, pp. 359–396. Marcel Dekker, New York (2004)
89. Van Ingen, H.E., Chan, D.W., Hubl, W., Miyachi, H., Molina, R., Pitzel, L., Ruibal, A., Rymer, J.C., Domke, I.: Analytical and clinical evaluation of an electrochemiluminescence immunoassay for the determination of CA 125. Clin. Chem. **44**, 2530–2536 (1998)
90. Xu, X., Jeffers, R. B., Gao, J., Logan, B.: Novel solution-phase immunoassays for molecular analysis of tumor markers. Analyst **126**, 1285–1292 (2001)
91. Yan, G., Xing, D., Tan, S., Chen, Q.: Rapid and sensitive immunomagnetic-electrochemiluminescent detection of p53 antibodies in human serum. J. Immunol. Methods **288**, 47–54 (2004)
92. Li, X., Wang, R., Zhang, X.: Electrochemiluminescence immunoassay at a nanoporous gold leaf electrode and using CdTe quantun dots as labels. Microchim Acta **172**, 285–290 (2011)
93. Li, C., Lin, J., Guo, Y. and Zhang, S. A novel electrochemiluminescent reagent of cyclometalated iridium complex-based DNA biosensor and its application in cancer cell detection. Chem. Comm. **47**, 4442–4444 (2011)
94. Jie, G., Wang, L., Zhang, S.: Magnetic electrochemiluminescent Fe3O4/CdSe–CdS nanoparticle/polyelectrolyte nanocomposite for highly efficient immunosensing of a cancer biomarker. Chem. Eur. J. **17**, 641–648 (2011)
95. US Patent: 0096918 A1 (2004)
96. Kurita, R., Arai, K., Nakamoto, K., Kato, D., Niwa, O.: Development of electrogenerated chemiluminescence-based enzyme linked immunosorbent assay for sub-pM detection. Anal. Chem. **82**, 1692–1697 (2011)

97. Xu, S., Liu, Y., Wang, T., Li, J.: Positive potential operation of a cathodic electrogenerated chemiluminescence immunosensor based on luminol and graphene for cancer biomarker detection. Anal. Chem **83**, 3817–3823 (2011)
98. Lin, D., Wu, J., Yan, F., Deng, S., Ju, H.: Ultrasensitive immunoassay of protein biomarker based on electrochemiluminescent quenching of quantum dots by hemin bio-bar-coded nanoparticle tags. Anal. Chem. **83**, 5214–5221 (2011)
99. Sardesai, N.P., Pan, S., Rusling, J.F.: Electrochemiluminescent immunosensor for detection of protein cancer biomarkers using carbon nanotube forests and $Ru(bpy)_3$-doped silica nanoparticles. Chem. Comm. **33**, 4968–4970 (2009)
100. Zu, Y., Bard, A.J.: Electrogenerated chemiluminescence. 67. Dependence of light emission of the tris(2,2′)bipyridylruthenium(II)/tripropylamine system on electrode surface hydrophobicity. Anal. Chem. **73**, 3960–3964 (2001)
101. Sardesai, N.P., Barron, J.C., Rusling, J.F.: Carbon nanotube microwell array for sensitive electrochemiluminescent detection of cancer biomarker proteins Anal. Chem. **83**, 6698–6703 (2011)
102. Yan, J.L., Estevez, M.C., Smith, J.E., Wang, K.M., He, X.X., Wang, L., Tan, W. H. Dye-doped nanoparticles for bioanalysis. Nano Today **2**, 44–50 (2007)
103. Knopp, D., Tang, D., Niessner, R.: Bioanalytical applications of biomolecule-functionalized nanometer-sized doped silica particles. Anal. Chim. Acta **647**, 14–30 (2009)
104. Choi, Y.E., Kwak, J.W., Park, J.W.: Nanotechnology for early cancer detection. Sensors **10**, 428–455 (2010)
105. Yang, X., Yuan, R., Chai, Y., Zhuo, Y., Mao, L., Yuan, S.: $Ru(bpy)_3$-doped silica nanoparticles labeling for a sandwich-type electrochemiluminescence immunosensor. Biosensors and Bioelectronics **25**, 1851–1855 (2010)
106. Qiana, J., Zhoub, Z., Caoa, X., Liu, S.: Electrochemiluminescence immunosensor for ultrasensitive detection of biomarker using $Ru(bpy)_3$-encapsulated silica nanosphere labels. Anal. Chim. Acta **665**, 32–38 (2010)
107. Bakalova, R., Zhelev, Z., Ohba, H., Baba, Y.: Quantum dot-based western blot technology for ultrasensitive detection of tracer proteins. J. Am. Chem. Soc. **127**, 9328–9329 (2005)
108. Jie, G.F., Liu, P., Zhang, S.S. Highly enhanced electrochemiluminescence of novel gold/silica/CdSe-CdS nanostructures for ultrasensitive immunoassay of protein tumor marker. Chem. Comm. **46**, 1323–1325 (2010)
109. Ding, Y., Kim Y. J., Erlebacher, J.: Nanoporous gold leaf: ancient technology/advanced material. Adv. Mater. **16**, 1897–1900 (2004)
110. Wei, F., Liao, W., Xu, Z., Yang, Y., Wong, D.T., Ho, C.-M.: Bio/abiotic interface constructed from nanoscale dna dendrimer and conducting polymer for ultrasensitive biomolecular diagnosis. Small **5**, 1784–1790 (2009)
111. Wei, F., Patel, P., Liao, W., Chaudhry, K., Zhang, L., Arellano-Garcia, M., Hu, S., Elashoff, D., Zhou, H., Shukla, S., Shah, F., Ho, C.-M., Wong, D.T.: Electrochemical sensor for multiplex biomarkers detection. Clin Cancer Res. **15**, 4446–4452 (2009)
112. Chikkaveeraiah, B.V., Mani, V., Patel, V., Gutkind, J.S., Rusling, J.F.: Microfluidic electrochemical immunoarray for ultrasensitive detection of two cancer biomarker proteins in serum. Biosens. Bioelectron. **26**, 4477–4483 (2011)
113. Jensen, G.C., Krause, C.E., Sotzing, G.A., Rusling, J.F.: Inkjet-printed gold nanoparticle electrochemical arrays on plastic. Application to immunodetection of a cancer biomarker protein. Phys. Chem. Chem. Phys. **13**, 4888–4894 (2011)
114. Tang, C.K., Vaze, A., Rusling, J.F.: Fabrication of immunosensor microwell arrays from gold compact discs for ultrasensitive detection of cancer biomarker proteins. Lab on a Chip (2011, in press). doi:10.1039/C1LC20833K
115. Cass, A., Davis, G., Francis, G.D., Hill, H.O.A., Aston, W.J., Higgins, I.J., Plotkin, E.V., Scott, L.D.L., Turner, A.P.F.: Ferrocene-mediated enzyme electrode for amperometric determination of glucose. Anal. Chem. **56**, 667–671 (1984)

Chapter 2
Nanomaterials for Biosensors and Implantable Biodevices

Roberto A. S. Luz, Rodrigo M. Iost and Frank N. Crespilho

Abstract The study of biological recognition elements and their specific functions has enabled the development of a new class of electrochemical modified electrodes called biosensors. Since the development of the first biosensor almost 50 years ago, biosensors technology have experienced a considerable growth in terms of applicability and complexity of devices. In the last decade this growth has been accelerated due the utilization of electrodes-modified nanostructured materials in order to increase the power detection of specific molecules. Other important feature can be associated with the development of new methodologies for biomolecules immobilization. This includes the utilization of several biological molecules such as enzymes, nucleotides, antigens, DNA, aminoacids and many others for biosensing. Moreover, the utilization of these biological molecules in conjunction with nanostructured materials opens the possibility to develop several types of biosensors such as nanostructured and miniaturized devices and implantable biosensors for real time monitoring. Based on recent strategies focused on nanomaterials for electrochemical biosensors development, these topics has presented recent methodologies and tools used until nowadays and the prospects for the future in the area.

R. A. S. Luz
Federal University of ABC (UFABC), Santo André, 09210-170, Brazil

R. M. Iost · F. N. Crespilho (✉)
Institute of Chemistry of São Carlos (IQSC), University of São Paulo (USP), São Carlos 13560-970, Brazil
e-mail: frankcrespilho@iqsc.usp.br

F. N. Crespilho (ed.), *Nanobioelectrochemistry*, DOI: 10.1007/978-3-642-29250-7_2,

2.1 Introduction

There is no doubt that the increase interest for the development of new materials applicable in electroanalytical techniques has been associated with the necessity of control specific molecules present in the environment or in more recently efforts, the human body [1–3]. This includes the possibility to improve the quality of life by development of efficient electrochemical devices and biodevices [2]. More than the use electrochemical devices to detect analytes is the challenge to develop more sensitive and selective electrochemical devices that provide the possibility to detect small quantities of molecules utilizing efficient transducing elements and specific recognition materials for biosensing [5–18]. The so called electrochemical biosensors are based on a specific biological recognition element such as enzymes, antigens or another biological molecule that interacts directly with a transducer element [19].

In general, a typical biosensor is composed for five parts (as illustrated in Fig. 2.1): (1) bioreceptors that bind of specific form to the analyte; (2) an electrochemically active interface where specific biological processes occur giving rise to a signal; (3) a transducer element that converts the specific biochemical reaction in an electrical signal that is amplified by a detector circuit using the appropriate reference; (4) a signal processor (e.g. computer software) for converting the electronic signal to a meaningful physical parameter describing the process being investigated and finally, (5) an proper interface to present the results to the human operator. Currently, the biosensors can be applied to a large variety of samples including body fluids, food samples, cell cultures and be used to analyze environmental samples [20].

The basic principles of electrochemical biosensors are associated with their capability to detect a specific molecule with high specificity. Also, these characteristics are dictated by a better correlation between the biological component and the transducing element. Important advances in these aspects has been achieved with the utilization of several kinds of nanomaterials such as metal nanoparticles [21], oxide nanoparticles [22], magnetic nanomaterials [23], carbon materials [24, 25] and metallophthalocyanines [26] to improve electrochemical signal of biocatalytic events occurred at electrode/electrolyte interface.

Recent advances in bionanoelectrochemistry are being reported about the enormous impact of nanomaterials when utilized as transducing element in modified electrodes [27–30]. Since then, thousands of scientific articles exploring the favorable association between biomolecules and nanomaterials to improve electrical signal originated in biochemical reactions have been published. One interesting example is the use of thin films on electrode surfaces to increase the sensitivity of sensors and biosensors. This sense, the Langmuir–Blodgett was the pioneering technique for the fabrication of thin films formed by transferring an amphiphilic material dispersed at air/water interface to a solid substrate. In particular, the obtention of thin monolayer films is very attractive for enzymes immobilization, proteins, nucleic acids and others [31]. Another interesting technique for fabrication

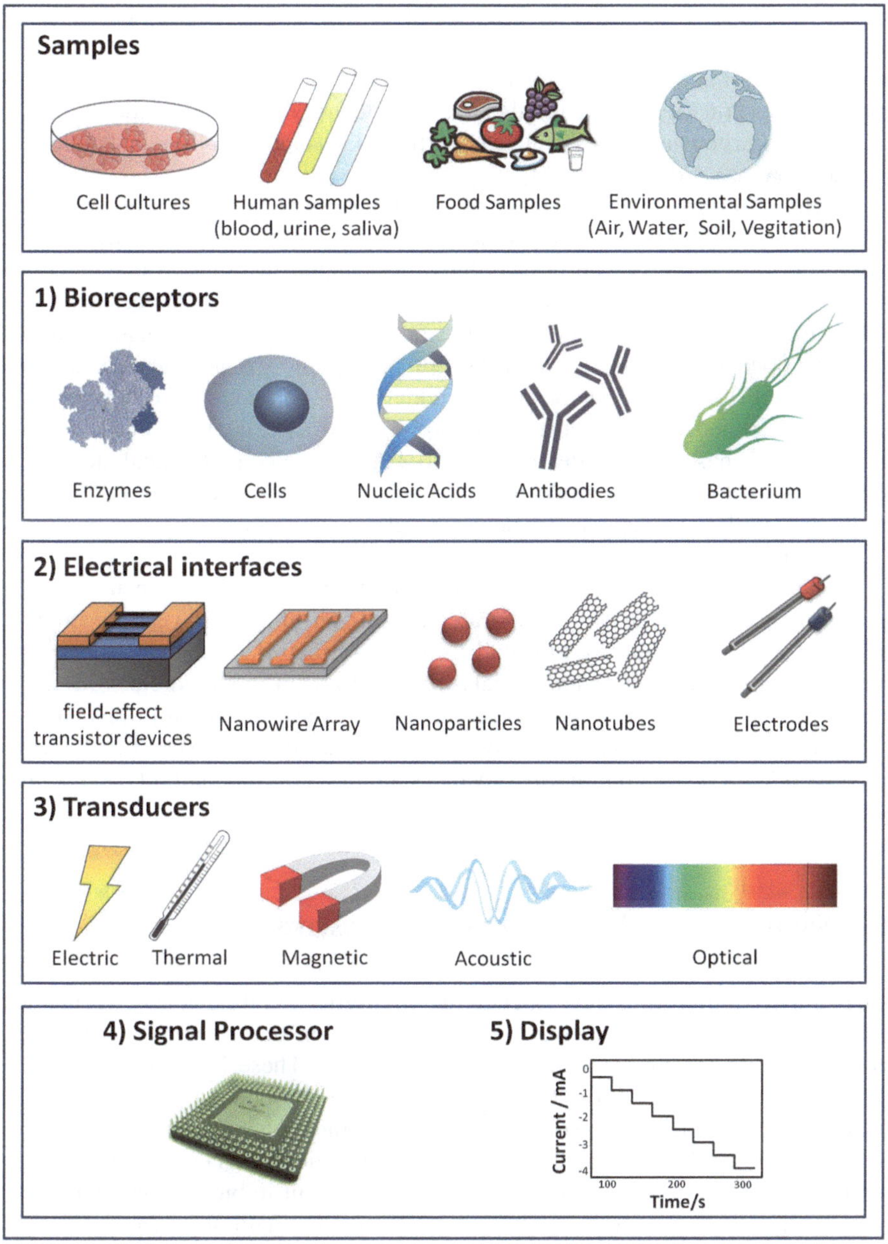

Fig. 2.1 Components of a typical biosensor

of thin organic films was developed by Decher in the beginning of 90 decade [32–34] as a simple strategy for fabrication of multilayer films with high control of thickness at nanoscale level. Instead of specific chemical interactions between substrate and organic molecules the layer-by-layer (LBL) technique is based, basically, in coulombic electrostatic that provides multilayers growth. Such strategy has been reported as an interesting tool for films fabrication with simplicity, which can be applied to several kinds of materials likes polyelectrolytes [35], metallophthalocyanines [36], carbon nanotubes [37], nanoparticles [38] and also biological molecules such as enzymes and proteins [39].

In what concerns the fabrication of electrochemical biosensors, it is unquestionable the importance of nanostructured materials and their implications in biosensors properties. Recent efforts have been made in order to use the nanostructured modified electrodes for monitoring specific biological molecules in vivo [40]. Also, the possibility to detect a specific molecule in living organisms at real time has open new paths for controlled of pathogenic diseases and, also, some analytes such as glucose at the human body [41]. Although these new class of electrodes opened the possibility to improve electrochemical biosensors performance, focus has been made in order to fabricate electrochemical devices at nanoscale level for single molecule detection [42]. Moreover, one of the main challenge until nowadays is to detect single events originated by enzymatic reactions utilizing a unique nanomaterial [43].

In this chapter, we describe the recent trends in the field of electrochemical biodevices exploring the principal strategies utilized in the last years to improve signal response of enzymatic biocatalysis and describe briefly the electrochemical characteristics of several nanomaterials when utilized in modified electrodes. The fabrication of nanoelectrodes by some techniques is also explored in this chapter. In addition, we will discuss the many efforts in order to detect specific molecules in vitro and in vivo and recent advances in the development of implantable biosensors.

2.2 Nanostructured Thin Films for Biosensing

Nanostructured thin films have opened the possibility to fabricate electrochemical sensors and biosensors with high power of detection due to intrinsic properties associated with their dimensions at nanoscale level. These interesting properties can be explained based on the organization level obtained when molecular arrangement is obtained at a solid conductor substrate. Also, the materials that can be used include a large range of organic and inorganic materials for films growth. Moreover, the possibility to improve the detection limit in biosensing devices can be also explained by using compatible materials such as natural polymers. The aim objective behind the utilization of these materials is to combine the high power of detection with preservation of the structural integrity of the biomolecules and, also, maintaining their biocatalytic activity.

2.2.1 *Langmuir–Blodgett and Layer-by-Layer Based Biosensors*

The field of materials science has opened new possibilities towards the utilization of organic, inorganic nanostructured materials and hybrids formed by biological components and nanostructured materials. In parallel, composites has been develop to confer or improve some specific properties which includes the use of metallic nanostructures or organic polymers. In particular, nanostructured organic films has opened a new research area with the aim purpose to obtain interesting properties at nanoscale. Nanostructured thin films has showed great impact in the field of electrochemical biosensors in the past few years with a large range of materials that can be employed in films construction. The study of organic molecules has arised since from 1960s decade with the discovery of their electronic properties and potential application in optic and electronic devices [44]. The major interest behind the utilization of nanostructured thin films for biosensing lies in the possibility to understand biochemical mechanisms and, at the same time, to fabricate mimetic systems based on cellular membranes [45]. The role of the control of depositing monolayers of organic films and their final properties was first studied by Irving Langmuir and Katherine Blodgett in the beginning of XX century [46, 47]. This technique of thin films fabrication is based on the self-organization of amphiphilic molecules at air/water interface in order to diminish the free surface energy and form a dispersed monolayer. The formation of organic monolayers is obtained by dropping of a dilute lipid solution at air/water interface with subsequent solvent evaporation. Also, the more stable monolayer conformation of Langmuir film formed on air/water interface is achieved by application of a horizontal and controlled compression throughout the Langmuir cube. Further, the compression is accomplished by two moves barriers localized at cube and is accompanied by measurement of certain surface properties such water surface tension and surface potential. The surface tension of water with the dispersion of an amphiphilic molecule on water interface can be measure utilizing Eq. 2.1.

$$\pi = \gamma_0 - \gamma \tag{2.1}$$

where π is the measurement of water surface tension change, γ_0 is the surface tension of pure water and γ is the surface tension of water with the presence of amphiphilic molecule at air/water interface. Although amphiphilic molecules are common used due to their self-organization at air-water interface, the dispersion of organic or inorganic molecules at interface is not considered to be limited to specific molecules. Moreover the type of substrate functionalization plays an important role for films formation. According to substrate functionalization, the monolayers can be transferred by immersion of substrate through the interface containing the amphiphilic monolayer. Consequently, the transfer of monolayers to the substrate is carried out by successive dipping the substrate in the cube. Also, the interaction during the substrate dipping is based on monolayers functionalization and the Langmuir films with X, Y and Z-type can be obtained [44]. One of

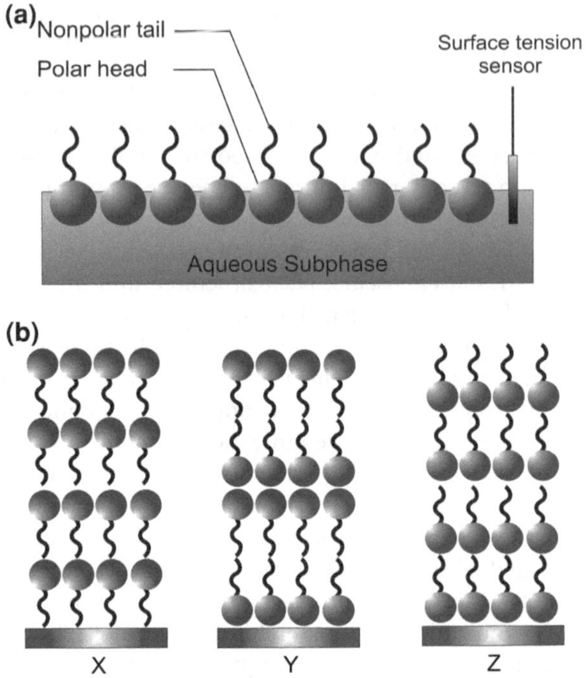

Fig. 2.2 **a** Schema for a Langmuir Monolayer obtained at air–water interface. **b** X, Y and Z Langmuir-Blodgett films obtained according to substrate and molecules used for films fabrication.

the major and interesting advantage is the possibility to control thickness and roughness by adsorption of multilayer films onto solid substrates. Figure 2.2 shows a schematic representation of a) Langmuir cube and b) the type of monolayer deposition according to the substrate functionalization and molecules used for films fabrication.

In the field of electrochemical biosensors, the utilization of biomolecules such as antibodies, DNA, enzymes or another kind of proteins adhered to Langmuir–Blodgett films confer specificity to the system [48–50]. Concerned the development of modified electrodes for enzymes immobilization, Langmuir–Blodgett films has been considered an important path for biosensors fabrication and many kinds of arquitectures has been reported in the last decades as very promissing approaches for biosensors development. Examples of biosensors development using LB method has been extensively reported on literature for application in several biosensing approaches [51, 52].

Several examples are reported about the determination of glucose using LB method as mimetic membrane platform for enzyme glucose oxidase (GOx) immobilization [18]. As an example, Sun and co-workers [53] reported the utilization of LB films for GOx immobilization utilizing cross-linking agents to improve biological process when enzyme was immobilized at monolayer surface. On another approach, Ohnuki and co-workers [54] reported the use of Langmuir films consisting of octadecyltrimethylammonium (ODTA) and Prussian blue (PB) clusters as platforms for enzyme GOx immobilization. The immobilization

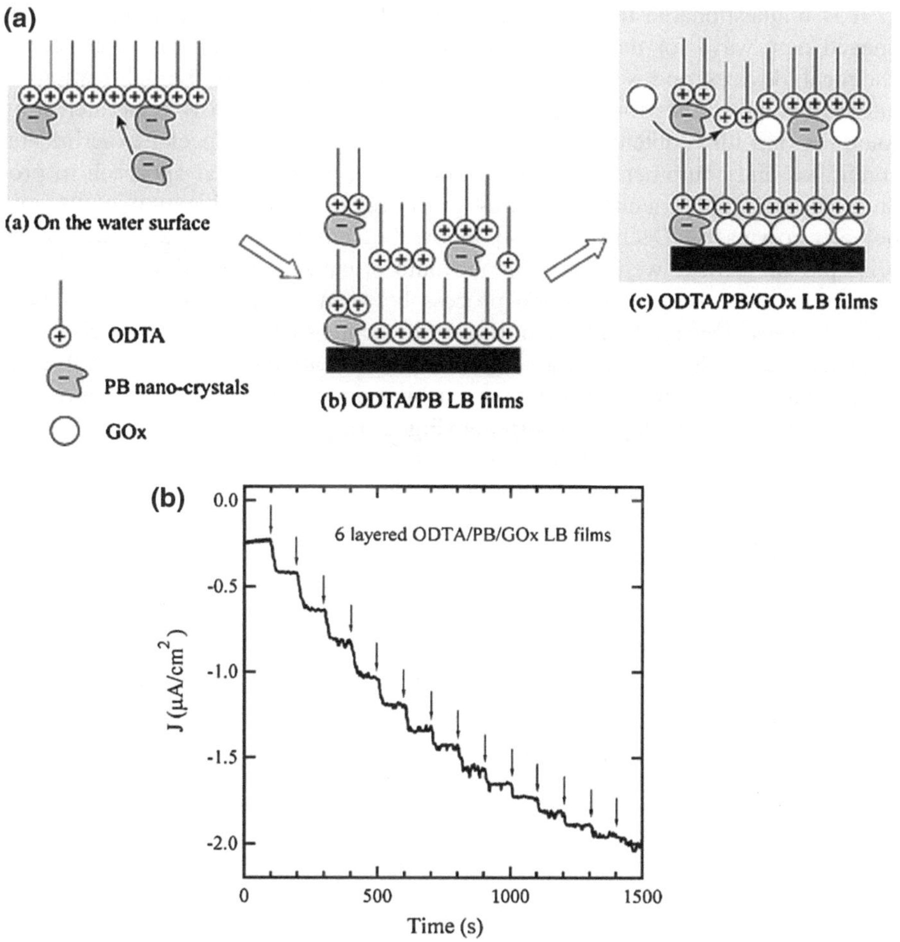

Fig. 2.3 **a** Scheme of LB films preparation containing ODTA, PB, and GOx. **b** Amperometric response obtained at 0.0 V in a buffer solution at pH 7.0 with ODTA/PB/GOx LB films (6 layers) deposited on a gold electrode. The arrows show the moment of glucose solution injection whose amount corresponds to an increase of 1 mmol L^{-1} glucose concentration. Reproduced with kind permission of Ref. [54]

of enzyme GOx was confirmed by FTIR spectra before and after enzyme immobilization with ODTA/PB Langmuir films. The configuration exhibited shows a good amperometric response upon glucose addition with the utilization of 6 layers, electrochemical increase process associated with the presence of PB electrocatalyst. Figure 2.3 shows a schematic representation of ODTA/PB Langmuir films and the amperometric response obtained at 0.0 V (Ag/AgCl).

It is unquestionable that the exploration of self-assembly methodologies has opened new ways for the development of more selective and sensitive electrochemical devices and so on the LB method has provided the fabrication of interesting approaches for biosensors development. Although LB is an interesting route for thin films obtention with high quality, it requires especial experimental conditions and equipments for films growth. The experimental approach to produce organic thin organic films was extended with the utilization of organic polyelectrolytes by Decher [32–34] in the beginning of 90 decade which the principle of growth were based in the auto-organized molecules primarily by coulombic electrostatic adsorption process between polyelectrolytes with oppositely charges. This method for multilayer films obtention is basically described by immersing a conducting substrate alternatively on cationic and anionic polyelectrolyte during a specific time which is washing on solvent solution to remove the excess between each step of adsorption (Fig. 2.4).

One important point about the formation of organic bilayers is the films stability achieved in the association of multilayers due to films growth. The energy of coulombic association is very low when each interaction of ions pairs is availed. However, the global association along polyelectrolyte interaction provides a high stability between polyelectrolytes chain [55]. The explanation for the association between polyelectrolyte multilayers were described utilizing the concept of extrinsic and intrinsic charges compensation when the association of two polyelectrolytes with opposite charges interacts [56]. The intrinsic charge compensation is described by the charge association of polyelectrolytes chains and is the basic explanation for the multilayers formation. At the same time, the charge balance of counter ions or the balance of extrinsic charges occurred with the polyelectrolyte chains during the formation of multilayers. For this purpose, some works described the polyelectrolyte association in terms of doping salt and the thermodynamic constant k_{dop} of extrinsic charge association, represented by Eq. 2.2 [57].

$$K_{dop} = \frac{y^2}{(1-y)a_{MA}^2} \approx \frac{y^2}{a_{MA}^2} \tag{2.2}$$

where y is the compensated fraction of polyelectrolyte charge and a_{MA} is the activity association between cation and anion. Much more details about the multilayers formation was described by several studies with the model of multilayers interpenetration between adjacent monolayers [58]. Jomaa and co-workers [59] utilized neutron reflectivity studies with interdispersed layers of deuterated poly(styrenesulfonate) (PSS) and poly(diallyldimethylammonium) (PDAC) to describe the formation of multilayer films. The exploration of films growth has also been described on literature by controlling several experimental conditions such as pH, concentration of salt or ionic strength, concentration of polyelectrolytes, temperature of the system, the solvent utilized, time of deposition, the nature of substrate and so on [58]. Also, the high control of film properties such as roughness, thickness and films stability can be obtained by controlling these conditions and plays an important role for the quality and stability of multilayers

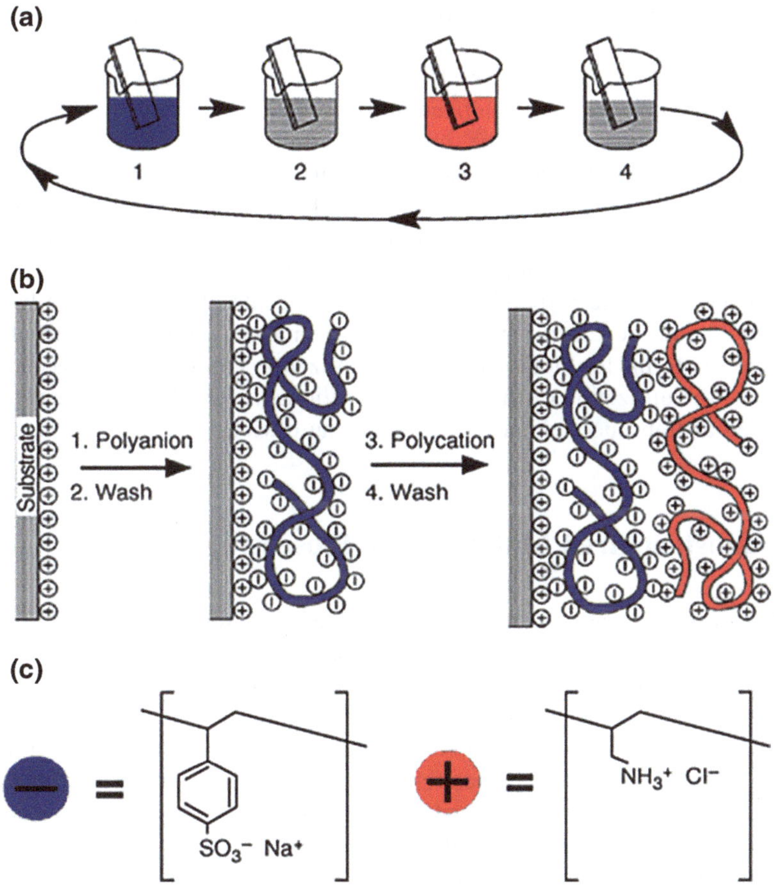

Fig. 2.4 **a** Schematic of the film deposition process using slides and beakers. Steps 1 and 3 represent the adsorption of a polyanion and polycation, respectively, and steps 2 and 4 are washing steps. The four steps are the basic buildup sequence for the simplest film architecture $(A/B)_n$. The construction of more complex film architectures requires only additional beakers and a different deposition sequence. **b** Simplified molecular picture of the first two adsorption steps, depicting film deposition starting with a positively charged substrate. Counterions are omitted for clarity. The polyion conformation and layer interpenetration are an idealization of the surface charge reversal with each adsorption step. **c** Chemical structures of two typical polyions, the sodium salt of poly(styrene sulfonate) and poly(allylamine hydrochloride). Reproduced with kind permission of Ref. [32]

achieved. Besides the possibility to obtain thin organic platforms with experimental simplicity compared to other techniques, LBL method has the advantage to incorporate a large range of materials in films fabrication that includes organic and inorganic materials, hybrids formed by materials at nanoscale and biological components. Regards the applicability of thin films for biosensing devices, the incorporation of biomolecules as components for films growth was described by

Lvov and co-workers [60] in the formation of self-organized multilayer films of proteins and polyelectrolytes of mioglobin (Mb) and PSS and by enzyme GOx and poly(ethylene imine) (PEI). Several other works reported the utilization of LBL method for biomolecules immobilization focused on biosensing applications. One important question is about the interaction study of nanomaterials and such biological molecules and their impact on biological process when biomolecules are exposed outer their natural environment [61]. The principal question is about the changes in molecular structure whose reflect directly on their biological properties. Moreover, the biological properties reflect directly on biosensors quantification and their capability to respond to a specific molecule. This properties has been achieved with the development and utilization of nanostructured thin films that can be acts as platforms for biomolecules immobilization [36, 62]. The next topic will emphasize the utilization of these hybrid functional materials and their capability to be used as transducer elements in biosensing devices.

2.3 Nanostructured Materials for Biosensing Devices

Nanostructured materials are well known as interesting tools with specific physical and chemical properties due to quantum-size effects and large surface area that provides unique and different properties compared to bulk materials. The exploration of these different characteristics provides the possibility to improve biosensors properties and increase the power of detection throughout size and morphology control. Interesting approaches has reported about the high increase in electronic properties when metallic nanostructures are used as components for electrodes modification. These includes the utilization of nanostructured materials with specific forms such 0D (quantum dots, nanoparticles), 1D (nanowires or carbon nanotubes) or 2D (metallic platelets or graphene sheets) orientation that reflects in their final properties. The next topic will be emphasize in biosensors fabrication using metallic nanoparticles (MNPs) as transducing elements on modified electrodes and some interesting electrochemical approaches used to improve biosensing performance.

2.3.1 Nanoparticles-Based Biosensors

Metallic nanoparticles are very interesting materials with unique electronic and electrocatalytic properties which depends on their size and morphology [63, 64]. The efficiency of electronic and electrochemical redox properties becomes these classes of nanostructured materials very interesting for technological applications. In particular, gold nanoparticles (AuNPs) are much explored materials as components for biosensors development due to the capability to increase electronic signal when a biological component is maintained in contact with nanostructured surface. On the other hand, silver, platinum, palladium, cooper, cobalt and others has extensively explored in biosensors development [65–69]. In particular, the exploration of gold nanostructured materials has provided new paths for enzymatic biosensors development. At the same time, specific organic stabilizers have been used to produce nanostructured materials with different morphologies. Dendrimers are known as organic macromolecules with tridimensional and highly defined structure functionality [70]. The capability of dendrimeric structures to stabilized and maintains integrity of metallic nanoparticles was reported by Crooks and co-workers [71]. As an example, polyamidoamine dendrimers (PAMAM) were used as template for nanoparticles growth or nano reactors with cavities for nanoparticles nucleation. According to functional groups at molecular structure, dendrimers has been subject of intense studies in the field of nanostructured thin films fabrication and also, in the form of hybrids with metallic nanoparticles. An interesting approach was reported recently utilizing hybrids of PAMAM-AuNPs as components in multilayer thin films based on LBL technique to enhance charge transfer in modified electrodes leading to the concept of electroactive nanostructured membranes (ENM) [72]. In this case, PAMAM-AuNP hybrids were assembled utilizing LBL technique in multilayers to produce modified electrodes. The strategy to produce modified substrates is based on self-assembly of polyvinylsulfonate (PVS) as negatively charged polyelectrolyte alternating with the positively charged PAMAM-AuNP hybrids onto ITO (indium tin oxide) conducting electrodes to obtain ENM. Also, the strategy involved the deposition of a redox mediator around metallic AuNPs to enhance charge transfer in modified electrodes (Fig. 2.5).

The capability to increase charge transfer utilizing the LBL approach was investigated with details by electrochemical impedance spectroscopy (EIS) was also evaluated using electrodeposition of different redox mediators (ITO-PVS/PAMAM-AuNP@Me). This approach can be generalized for a wide range of electrochemical devices, including sensors and biosensors. The enhanced charge transport on electrodes based on LBL approach was also explored by electrodeposition of Prussian blue redox mediator (PB) on PAMAM-AuNP nanocomposite [73]. The electrochemical results shows kinetic behavior correlation for cathodic current peak for AuNPs showed a non-linear response compared to adsorption time for bilayers formation.

2.3.2 Carbon Materials-Based Biosensors

Carbon materials have received great attention in the last decades with the emergence of nanoscience area [75]. The utilization of carbon nanomaterials also possibilities the increase on charge transfer in bioelectrochemical devices. These includes the modification of electrodes with several kinds of carbon at nanometer range carbon powder, carbon nanotubes, graphene sheets and carbon capsules [76–78]. The investigation of electronic properties of carbon nanotubes since their discovery by Iijima and co-workers [79] in 1991 are one of the most reported

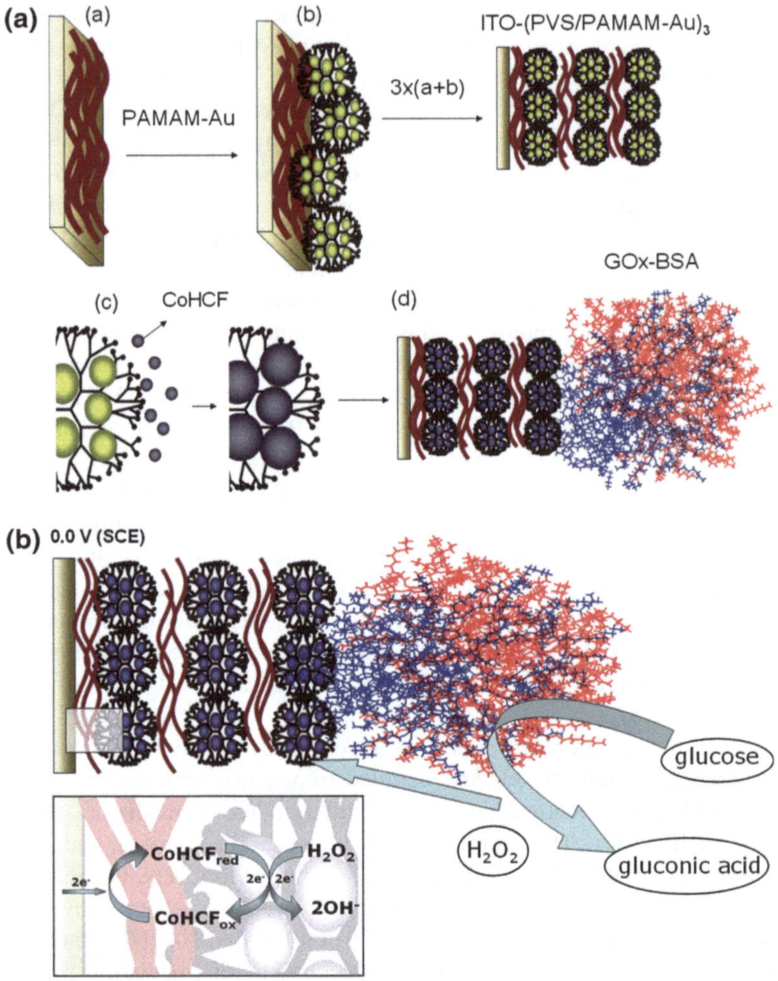

Fig. 2.5 **a** Schematic fabrication of LbL films comprising PVS and PAMAM-Au. The sequential deposition of LbL multilayers was carried out by immersing the substrates alternately into (**a**) PVS (**a**) and PAMAM-Au (**b**) solutions for 5 min per step. After deposition of 3 layers, an ITO-(PVS/PAMAM-Au)$_3$@CoHCF electrode was prepared by potential cycling (**c**). The enzyme immobilization to produce ITO-(PVS/PAMAM-Au)$_3$@CoHCF-GOx (**d**) was carried out in a solution containing BSA, glutaraldehyde and GOx. **b** Schematic representation of reaction of glucose at ITO-(PVS/PAMAM-Au)$_3$@CoHCF-GOx electrode. Reprinted with permission from Ref. [74] Copyright 2007 Elsevier

approaches used to explain their capability to increase the detection limit in modified electrodes. The intrinsic electronic properties of carbon materials can be explained based on the nature of carbon bonding in their allotropic forms. Graphite is the simplest form of carbon–carbon bond with sp^2 hybridization with weak bond energy between adjacent layers and σ bond with and out of plane of π orbitals.

Carbon nanotubes (CNT) are formed by a hollow cylinder formed by a unique carbon sheet forming a single walled carbon nanotube (SWCNT) or concentric carbon sheets with different diameters forming multiwalled carbon nanotubes (MWCNT) with carbon–carbon with sp^2 bonding [75]. The particular cylindrical form of CNT is the principal aspect that provides the quantum confinement effect in the oriented 1D nanostructured materials [80]. These characteristics provide the possibility to increase chemical reactivity and electronic properties of this particular carbon material, which becomes a crucial point for biosensing devices [75].

The electrochemical properties of CNTs have also been considered an interesting point for biosensors fabrication. Initially, edge-planes sites and defect areas present on tubes structures has been the focus of intense studies about their electroactivity. On the other hand, some interesting studies reported about iron impurity present on CNTs and their influence on electrocatalytic activity [81]. Anodic or cathodic pre-treatments have also been employed principally for detection of biological systems. Liu and co-workers [82] reported the preparation of PDDA/GOx/PDDA/CNT-modified glassy carbon electrode self-assembly nanocomposite for flow injection glucose biosensing applications. The modified electrodes was obtained through electrostatic adsorption between adjacent bilayers (LBL method) showing linear response at range of 15 μM to 6 mM and detection limit of 7 μmol L^{-1} for H_2O_2. Another interesting approach was reported recently using nanoarquitectures based on capacitive field effect transistor-modified with LBL of PAMAM and CNTs as sensing platforms for penicillin G detection [14]. The large surface area provide by incorporation of a organic matrix and the increased response with CNT incorporation on the modified transistors exhibited and excellent and faster response upon addition of penicillin on electrolytic media. The same electrode configuration based on PAMAM/CNT arquitectures has also provided the improvement of biosensing effects for glucose biosensor ranges from 4.0 μmol L^{-1} to 1.2 mM and limit detection of 2.5 μmol L^{-1}. Other applications include the utilization of several other arquitectures (Fig. 2.6a) with interesting electrochemical properties upon immobilization of biomolecules such as DNA [83] and antigens for immunosensing applications [84]. Li and co-workers [85] reported the utilization of MWCNT nanoelectrodes highly oriented embedded on SiO_2 for ultrasensitive DNA detection. The combine of redox species $Ru(bpy)_3^{2+}$ mediated guanine oxidation possibilities detection of small quantities of redox substances-based immunosensing applications (Fig. 2.6b).

Graphene sheets (GS) have recently attracted much attention in the field for electrochemical sensing and biosensing areas [86, 87]. The 2D electronic structure of graphene was investigated in detail in several articles due to their potential application as components in a large range of electrochemical devices [88]. The aim advantage of GS is their large surface area when compared to CNTs and consequently, their electrochemical properties can increase enormous when biological molecules are immobilized on electrode surface [25]. One interesting method for synthesis of graphene conductor sheets is based on (chemical, physical or electrochemical preparation) insulator graphene oxides as precursor to form graphene conductor structures. Several studies emphasized changes in structural

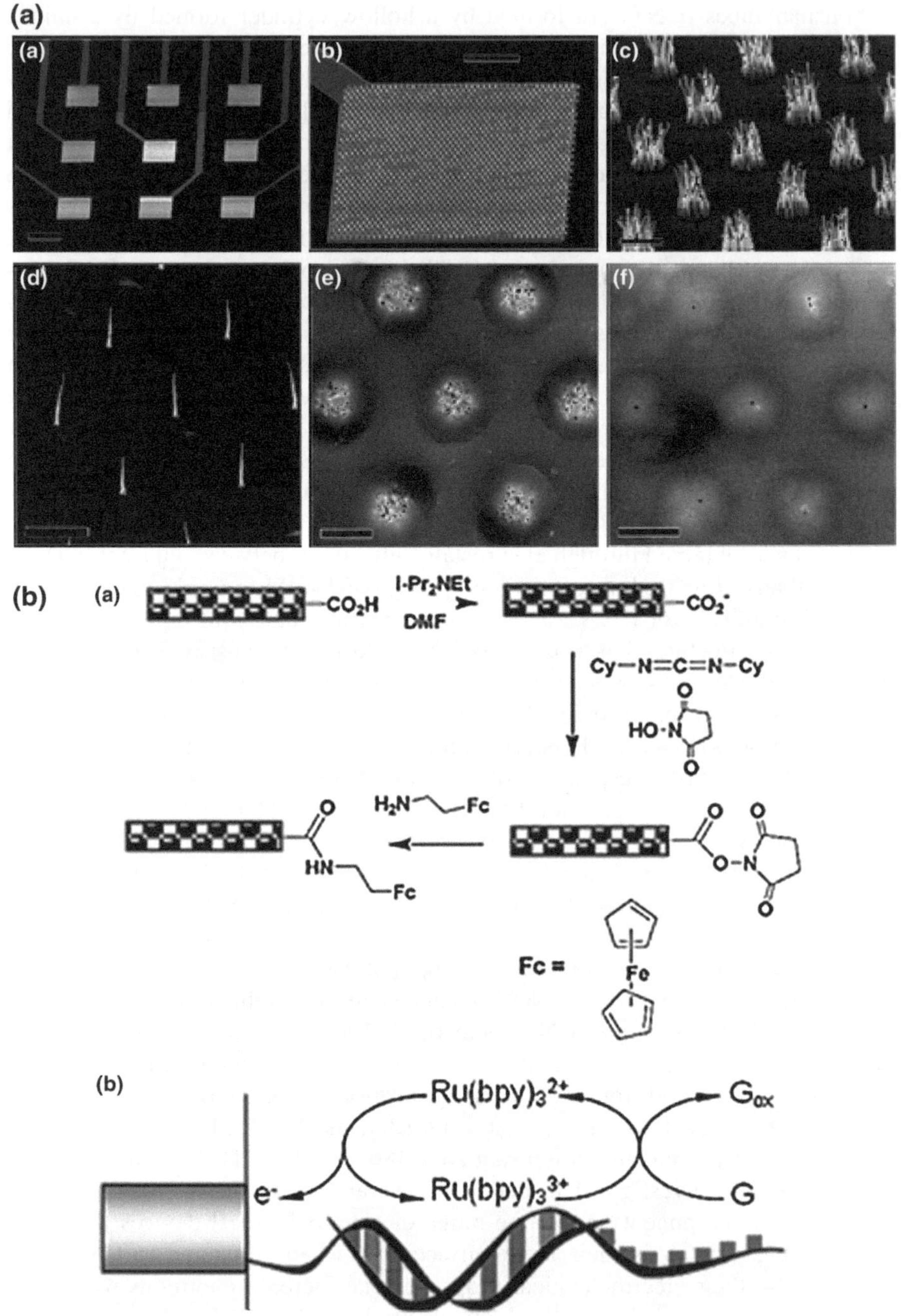

Fig. 2.6 **a** SEM images of (**a**) 3 × 3 electrode array, (**b**) array of MWNT bundles on one of the electrode pads, (**c**) and (**d**) array of MWNTs at UV-lithography and e-beam patterned Ni spots, respectively, (**e**) and (**f**) the surface of polished MWNT array electrodes grown on 2 ím and 200 nm spots, respectively. Panels (**a–d**) are 45° perspective views and panels (**b–f**) are top views. The scale bars are 200, 50, 2, 5, 2, and 2 ím, respectively. (**b**) (**a**) The Functionalization Process of the Amine- Terminated Ferrocene Derivative to CNT Ends by Carbodiimide Chemistry and (**b**) the Schematic Mechanism of $Ru(bpy)_3^{2+}$ Mediated Guanine Oxidation [85]

properties of graphene according to the methodology employed in their fabrication [89]. Shan and co-workers [90] studied the influence of polyvinylpyrrolidone–protected graphene/polyethylenimine–functionalized ionic liquid/GOx in modified electrodes for glucose biosensing. This electrode configuration showed high electrochemical sensibility and biocompatibility when enzyme GOx was immobilized at electrode surface. These two combined properties of biocompatibility and improvement of electrochemical sensibility upon addition of H_2O_2 and O_2 in electrolytic media shows their potential application in biosensor devices. On another interesting approach, Kang and co-workers [91] reported about the utilization of nanocomposites based on GS and chitosan (Ch) organic natural polymer as platforms for glucose sensing. It is well known that Ch is a natural polymer that provides the ability to improve electrochemical redox process when used in modified electrodes. Also, this electrochemical approach has been much studied as promising methods for enzyme immobilization, as in the case of enzymatic biosensors development with sensitivity of 37.93 $\mu A\ mM^{-1}\ cm^{-1}$ at linear range of 0.08 mmol L^{-1} to 12 mmol L^{-1}.

Although carbon-based nanostructured materials are relatively a recent area, their impact in the field of biosensors development has been arised significantly in the last decades as interesting approaches for biomolecules study. Such progress can be attributed to the intense research in nanocomposites development applicable in electrochemistry devices with unique electronic properties. These features includes the utilization of different carbon materials such CNT and GS as platforms for enhance electronic signal between electrode surface and biomolecules such as oxidoreductases enzymes. Concerning the development of more sensitive biosensing devices, the crescent use of nanostructured materials for improve electronic communication of biological materials and electrode surface plays an important role for detection of small quantities of molecular substances.

2.4 Miniaturized Devices and Implantable Biosensors

Besides the modification of electrodes surface by nanomaterials, in recent years, some studies have been done in trying to build biosensors and bioelectronics devices with nanometric geometry [42, 92], where the individual 1D structures are applied as working electrodes for current measurements low, typically on the order of fentoamperes (f) and picoamperes (pA). Several types of electrodes such as single-walled carbon nanotubes (SWNTs) [93, 94], boron-doped silicon nanowires (SiNWs) [92] and Sn doped In_2O_3 nanowires (ITO-NWs) have been shown to be interesting for building nanodevices [42]. For example, in a pioneer work, Lemay and co-workers [95] performed electrochemical measurements, on reduced scale of redox enzymes to study a small amount of molecules. This approach was based on lithographically fabricated Au nanoelectrodes with dimensions down to ca. 70 × 70 nm, where was demonstrated successfully for the first time a distinct catalytic response from less than 50 enzymes ([NiFe]-hydrogenase) molecules. These results were obtained using cyclic voltammetry in which were observed a turnover current of 22 fA. However, because of high surface-to-volume ratio and tunable electron transport properties related to the quantum confinement effect present in these nanodevices, their electrical properties are strongly influenced by minor perturbations. This way, when an electrode with nanometer dimensions is used, various types of noises can affect the measurements and compromise the interpretation of the results.

Recently, the noise and distortions are the main factors limiting the accuracy of measurements in devices at low current conditions (sub-pico-Ampere). In experiments using electrodes macro-scale (centimeters, micrometers) problems related noises can be easily overcome by the use of programs for signal smoothing. However, for nanoelectrodes, the use of conventional methods of smoothing of signals can lead to loss of useful information. Thus, many research efforts have been observed in the development of methodologies capable of minimizing the effects of external disturbances in the low currents measurements in nanoelectrodes. In a pioneering study, Goncalves and co-workers [92] reported the development of numerical methods for smoothing signal and noise modeling. Like most of the noise frequency affecting the measurements are known (thermal, flicker, burst and shot noise) smoothing filters were used to promote a better visualization of the useful signal. Numerical methods have proven useful for the treatment of the signal due to its simplicity and speed of processing, allowing the identification of unwanted signals, changes in control parameters related to the final quality of the processed signal and quick view of the desired signal [92].

The miniaturization of electrochemical platforms is an important feature in the development of the new generation of implantable clinical devices for monitoring metabolites at living organisms [96]. The implantable biosensors are presented as ideally devices desirable for the diagnosis and management of metabolic diseases such as, diabetes, which currently is based on data obtained from test strips using drops of blood. Although widely used, this procedure is unable to reflect the general situation of the patient and point out trends and patterns associated with their daily habits. Thus, many studies focused on the development of implantable biosensors for continuous monitoring of several biologically important metabolites have been reported in bioelectrochemical area with the purpose to improve human quality of life and too in recent trends, the capability to generate energy from biomass fuels [97–99]. Figure 2.7, for example, shows a catheter microchip that consists of flexible carbon fiber electrodes modified with neutral red redox mediator (FTCF-NR) being implanted in jugular vein of rat. This system can be used both to monitor glucose levels and for power generation in biofuel cells utilizing enzymes and microorganisms.

Despite promising, the reliability of implantable systems is often undermined by factors like biofouling [100, 101] and foreign body response [102] in addition to sensor drifts and lack of temporal resolution [103]. To minimize such problems,

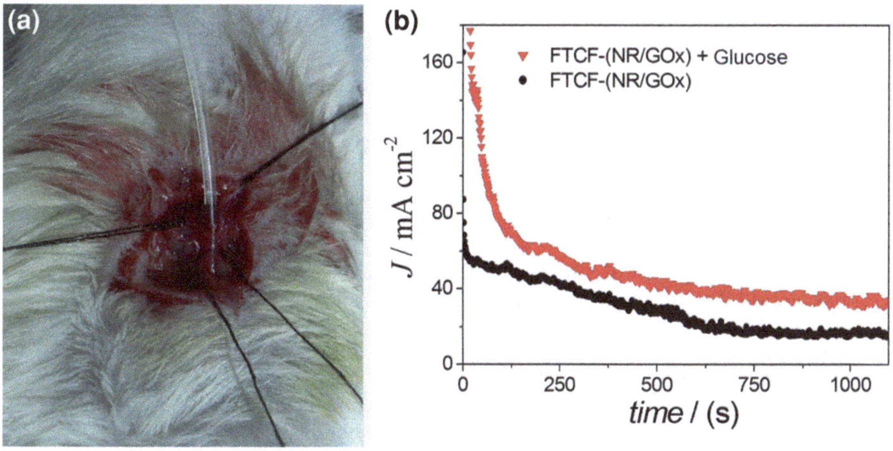

Fig. 2.7 **a** Photograph of implanted catheter microchips in jugular vein of rat from *Rattus Novergicus* species **b** Chronoamperometry curves in situ without the addition of glucose (*black line*) and with addition of glucose (*red line*)

many researchers have directed their work for the synergism between biosensors and nantotecnologia which has led to diagnostic devices more reliable [104, 105].

The prospects of implantable devices and in particular the metabolic monitoring can only be achieved if they can be readily implanted and explanted without the need for complicated surgery. this sense, for facilitated the implantation, the implantable device should be extremely small, which calls for miniaturization of various functional components, such as electrodes, power sources, signal processing units and sensory elements. This way, miniaturized biosensors can cause less tissue damage and therefore less inflammation and foreign body response [106].

2.5 Conclusion

Currently, research in the area of biosensing is conducted not only in the construction of miniaturized devices, faster, cheaper and more efficient, but also in the increasing integration of electronic and biological systems. This way, the future development of biosensors and devices bioelectronics analysis of highly sensitive and specific will require the combination multidisciplinary areas like quantum, chemistry and solid state physics and surface, bioengineering, biology and medicine, electrical engineering, among others. Advances in any of these fields will have significant effects on the future of medical diagnosis and treatment, where the monitoring continuous diseases, prevention methods and development of more effective drugs with side effects minimized will benefited by biosensing technologies.

References

1. Clark Jr, L.C., Lyons, C.: Electrode systems for continuous monitoring in cardiovascular surgery. Ann. N. Y. Acad. Sci. **102**(1), 29–45 (1962)
2. Castillo, J., et al.: Biosensors for life quality: design, development and applications. Sens Actuators B: Chem **102**(2), 179–194 (2004)
3. Suaud-Chagny, M.F.: In vivo monitoring of dopamine overflow in the central nervous system by amperometric techniques combined with carbon fibre electrodes. Methods **33**(4), 322–329 (2004)
4. Voskerician, G., Liu, C.C., Anderson, J.M.: Electrochemical characterization and in vivo biocompatibility of a thick-film printed sensor for continuous in vivo monitoring. Sens. J. IEEE **5**(6), 1147–1158 (2005)
5. Zhao, W., Xu, J.J., Chen, H.Y.: Electrochemical Biosensors Based on Layer by Layer Assemblies. Electroanalysis **18**(18), 1737–1748 (2006)
6. Singhal, R., et al.: Immobilization of glucose oxidase onto Langmuir-Blodgett films of poly-3-hexylthiophene. Curr. Appl. Phys. **3**(2–3), 275–279 (2003)
7. Fu, Y., et al.: One pot preparation of polymer–enzyme–metallic nanoparticle composite films for high performance biosensing of glucose and galactose. Adv. Funct. Mater. **19**(11), 1784–1791 (2009)
8. Wang, J., et al.: Glucose oxidase entrapped in polypyrrole on high-surface-area Pt electrodes: a model platform for sensitive electroenzymatic biosensors. J. Electroanal. Chem. **575**(1), 139–146 (2005)
9. Lin, J., et al.: One-step synthesis of silver nanoparticles/carbon nanotubes/chitosan film and its application in glucose biosensor. Sens. Actuators B: Chem. **137**(2), 768–773 (2009)
10. Wu, B.Y., et al.: Amperometric glucose biosensor based on multilayer films via layer-by-layer self-assembly of multi-wall carbon nanotubes, gold nanoparticles and glucose oxidase on the Pt electrode. Biosens. Bioelectron. **22**(12), 2854–2860 (2007)
11. Vamvakaki, V., Chaniotakis, N.A.: Carbon nanostructures as transducers in biosensors. Sens. Actuators B: Chem. **126**(1), 193–197 (2007)
12. Zhang, Y., et al.: Carbon nanotubes and glucose oxidase bionanocomposite bridged by ionic liquid-like unit: Preparation and electrochemical properties. Biosens. Bioelectron. **23**(3), 438–443 (2007)
13. Gooding, J.J., Hibbert, D.B.: The application of alkanethiol self-assembled monolayers to enzyme electrodes. TrAC Trends Anal. Chem. **18**(8), 525–533 (1999)
14. Siqueira Jr, J.R., et al.: Penicillin biosensor based on a capacitive field-effect structure functionalized with a dendrimer/carbon nanotube multilayer. Biosens. Bioelectron. **25**(2), 497–501 (2009)
15. Zhu, L., et al.: Bienzymatic glucose biosensor based on co-immobilization of peroxidase and glucose oxidase on a carbon nanotubes electrode. Biosens. Bioelectron. **23**(4), 528–535 (2007)
16. Deng, S., et al.: A glucose biosensor based on direct electrochemistry of glucose oxidase immobilized on nitrogen-doped carbon nanotubes. Biosens. Bioelectron. **25**(2), 373–377 (2009)
17. Abel, P.U., Von Woedtke, T.: Biosensors for in vivo glucose measurement: can we cross the experimental stage. Biosens. Bioelectron. **17**(11–12), 1059–1070 (2002)
18. Caseli, L.: Control of catalytic activity of glucose oxidase in layer-by-layer films of chitosan and glucose oxidase. Mater. Sci. Eng. **27**(5–8), 1108–1110 (2007)
19. Davis, F., Higson, S.P.J.: Structured thin films as functional components within biosensors. Biosens. Bioelectron. **21**(1), 1–20 (2005)
20. Grieshaber, D., et al.: Electrochemical biosensors-Sensor principles and architectures. Sensors **8**(3), 1400–1458 (2008)
21. He, B., Morrow, T.J., Keating, C.D.: Nanowire sensors for multiplexed detection of biomolecules. Curr. opin. Chem. Biol. **12**(5), 522–528 (2008)

22. De Dios, A.S., Díaz-García, M.E.: Multifunctional nanoparticles: Analytical prospects. Analytica Chimica Acta **666**(1–2), 1–22 (2010)
23. Haun, J.B., et al.: Magnetic nanoparticle biosensors. Wiley Interdiscip. Rev.: Nanomed. Nanobiotechnol. **2**(3), 291–304 (2010)
24. Rusling, J.F., Sotzing, G., Papadimitrakopoulosa, F.: Designing nanomaterial-enhanced electrochemical immunosensors for cancer biomarker proteins. Bioelectrochemistry **76**(1–2), 189–194 (2009)
25. Kim, Y.R., et al.: Electrochemical detection of dopamine in the presence of ascorbic acid using graphene modified electrodes. Biosens. Bioelectron. **25**(10), 2366–2369 (2010)
26. Sergeyeva, T.A., et al.: Hydrogen peroxide-sensitive enzyme sensor based on phthalocyanine thin film. Analytica Chimica Acta **391**(3), 289–297 (1999)
27. Yamauchi, F., et al.: Layer-by-layer assembly of cationic lipid and plasmid DNA onto gold surface for stent-assisted gene transfer. Biomaterials **27**(18), 3497–3504 (2006)
28. Iost, R.M., et al.: Strategies of Nano-Manipulation for Application in Electrochemical Biosensors. Int. J. Electrochem. Sci **6**, 2965–2997 (2011)
29. Musameh, M., et al.: Low-potential stable NADH detection at carbon-nanotube-modified glassy carbon electrodes. Electrochem. Commun. **4**(10), 743–746 (2002)
30. Dequaire, M., Degrand, C., Limoges, B.: An electrochemical metalloimmunoassay based on a colloidal gold label. Anal. Chem. **72**(22), 5521–5528 (2000)
31. Iost, R.M., et al.: Recent advances in nano-based electrochemical biosensors: application in diagnosis and monitoring of diseases. Frontiers Biosci. (Elite edition) **3**, 663 (2011)
32. Decher, G.: Fuzzy nanoassemblies: toward layered polymeric multicomposites. Science **277**(5330), 1232 (1997)
33. Decher, G., et al.: Layer-by-layer assembled multicomposite films. Curr. opin. Colloid & Interface Sci. **3**(1), 32–39 (1998)
34. Decher, G., et al.: New nanocomposite films for biosensors: layer-by-layer adsorbed films of polyelectrolytes, proteins or DNA. Biosens. Bioelectron. **9**(9–10), 677–684 (1994)
35. Bricaud, Q., et al.: Energy transfer between conjugated polyelectrolytes in layer-by-layer assembled films. Langmuir **27**(8), 5021–5028 (2011)
36. Alencar, W.S., et al.: Influence of film architecture on the charge-transfer reactions of metallophthalocyanine layer-by-layer films. J. Phys. Chem. C **111**(34), 12817–12821 (2007)
37. Luz, R.A.S., et al.: Supramolecular architectures in layer-by-layer films of single-walled carbon nanotubes, chitosan and cobalt (II) phthalocyanine. Mater. Chem. Phys. **130**(3), 1072–1077 (2011)
38. Alencar, W.S., et al.: Synergistic interaction between gold nanoparticles and nickel phthalocyanine in layer-by-layer (LbL) films: evidence of constitutional dynamic chemistry (CDC). Phys. Chem. Chem. Phys. **11**(25), 5086–5091 (2009)
39. Siqueira Jr, J.R., et al.: Immobilization of biomolecules on nanostructured films for biosensing. Biosens. Bioelectron. **25**(6), 1254–1263 (2010)
40. Njagi, J., et al.: Amperometric detection of dopamine in vivo with an enzyme based carbon fiber microbiosensor. Anal. Chem. **82**(3), 989–996 (2010)
41. Ward, W.K., et al.: A new amperometric glucose microsensor: in vitro and short-term in vivo evaluation. Biosens. Bioelectron. **17**(3), 181–189 (2002)
42. Crespilho, F.N., et al.: Development of individual semiconductor nanowire for bioelectrochemical device at low overpotential conditions. Electrochem. Commun. **11**(9), 1744–1747 (2009)
43. Besteman, K., et al.: Enzyme-coated carbon nanotubes as single-molecule biosensors. Nano Lett. **3**(6), 727–730 (2003)
44. Paterno, L.G., Mattoso, L.H.C., Oliveira Jr, O.N.: Filmes poliméricos ultrafinos produzidos pela técnica de automontagem: preparação, propriedades e aplicações. Quim. Nova **24**(2), 228–235 (2001)
45. Goto, T.E., et al.: Enzyme activity of catalase immobilized in Langmuir- Blodgett films of phospholipids. Langmuir,**26**(13), 11135–11139 (2010)

46. Blodgett, K.B.: Monomolecular films of fatty acids on glass. J. Am. Chem. Soc. **56**(2), 495–495 (1934)
47. Blodgett, K.B.: Films built by depositing successive monomolecular layers on a solid surface. J. Am. Chem. Soc. **57**(6), 1007–1022 (1935)
48. Santos, T.C.F., et al.: Mixing alternating copolymers containing fluorenyl groups with phospholipids to obtain Langmuir and Langmuir- Blodgett films. Langmuir **26**(8), 5869–5875 (2009)
49. Montanha, E.A., et al.: Properties of lipophilic nucleoside monolayers at the air-water interface. Colloids Surf. B: Biointerfaces **77**(2), 161–165 (2010)
50. Pavinatto, F.J., et al.: Cholesterol mediates chitosan activity on phospholipid monolayers and Langmuir- Blodgett films. Langmuir **25**(17), 10051–10061 (2009)
51. Schmidt, T.F., et al.: Enhanced activity of horseradish peroxidase in Langmuir-Blodgett films of phospholipids. Biochimica et Biophysica Acta (BBA)-Biomembranes **1778**(10), 2291–2297 (2008)
52. Caseli, L., et al.: Flexibility of the triblock copolymers modulating their penetration and expulsion mechanism in Langmuir monolayers of dihexadecyl phosphoric acid. Colloids Surf. B: Biointerfaces **22**(4), 309–321 (2001)
53. Sun, F., et al.: Polymer ultrathin films by self-assembly: bound perfluorinated monolayers and their modification using in situ derivatization strategies. Thin Solid Films **242**(1–2), 106–111 (1994)
54. Ohnuki, H., et al.: Immobilization of glucose oxidase in Langmuir-Blodgett films containing Prussian blue nano-clusters. Thin solid films **516**(24), 8860–8864 (2008)
55. Schlenoff, J.B., Rmaile, A.H., Bucur, C.B.: Hydration contributions to association in polyelectrolyte multilayers and complexes: visualizing hydrophobicity. J. Am. Chem. Soc. **130**(41), 13589–13597 (2008)
56. Clark, S.L., Hammond, P.T.: The role of secondary interactions in selective electrostatic multilayer deposition. Langmuir **16**(26), 10206–10214 (2000)
57. Bucur, C.B., Sui, Z., Schlenoff, J.B.: Ideal mixing in polyelectrolyte complexes and multilayers: entropy driven assembly. J. Am. Chem. Soc. **128**(42), 13690–13691 (2006)
58. Dubas, S.T., Schlenoff, J.B.: Factors controlling the growth of polyelectrolyte multilayers. Macromolecules **32**(24), 8153–8160 (1999)
59. Jomaa, H.W., Schlenoff, J.B.: Salt-induced polyelectrolyte interdiffusion in multilayered films: A neutron reflectivity study. Macromolecules **38**(20), 8473–8480 (2005)
60. Lvov, Y., et al.: Thin film nanofabrication via layer-by-layer adsorption of tubule halloysite, spherical silica, proteins and polycations. Colloids Surf A: Physicochem. Eng. Aspects **198**, 375–382 (2002)
61. Crespilho, F.N., et al.: Enzyme immobilization on Ag nanoparticles/polyaniline nanocomposites. Biosens. Bioelectron. **24**(10), 3073–3077 (2009)
62. Crespilho, F.N., et al.: Electrochemistry of Layer-by-Layer Films: a review. Int. J. Electrochem. Sci **1**, 194–214 (2006)
63. Park, K.W., et al.: Chemical and electronic effects of Ni in Pt/Ni and Pt/Ru/Ni alloy nanoparticles in methanol electrooxidation. J. Phys. Chem. B **106**(8), 1869–1877 (2002)
64. El-Deab, M.S., Ohsaka, T.: An extraordinary electrocatalytic reduction of oxygen on gold nanoparticles-electrodeposited gold electrodes* 1. Electrochem. Commun. **4**(4), 288–292 (2002)
65. Hrapovic, S., et al.: Electrochemical biosensing platforms using platinum nanoparticles and carbon nanotubes. Anal. Chem. **76**(4), 1083–1088 (2004)
66. Liu, C.Y., Hu, J.M.: Hydrogen peroxide biosensor based on the direct electrochemistry of myoglobin immobilized on silver nanoparticles doped carbon nanotubes film. Biosens. Bioelectron. **24**(7), 2149–2154 (2009)
67. Li, Z., et al.: Application of hydrophobic palladium nanoparticles for the development of electrochemical glucose biosensor. Biosens. Bioelectron. (2011)
68. Baioni, A.P., et al.: Copper hexacyanoferrate nanoparticles modified electrodes: A versatile tool for biosensors. J. Electroanal. Chem. **622**(2), 219–224 (2008)

69. Salimi, A., Hallaj, R., Soltanian, S.: Fabrication of a sensitive Cholesterol Biosensor based on cobalt oxide nanostructures electrodeposited onto glassy carbon electrode. Electroanalysis **21**(24), 2693–2700 (2009)
70. Astruc, D., Chardac, F.: Dendritic catalysts and dendrimers in catalysis. Chem. Rev. **101**(9), 2991–3024 (2001)
71. Crooks, R.M., et al.: Dendrimer-encapsulated metal nanoparticles: synthesis, characterization, and applications to catalysis. Acc. Chem. Res. **34**(3), 181–190 (2001)
72. Siqueira Jr, J.R., et al.: Bifunctional electroactive nanostructured membranes. Electrochem. Commun. **9**(11), 2676–2680 (2007)
73. Crespilho, F.N., et al.: Enhanced charge transport and incorporation of redox mediators in layer-by-layer films containing PAMAM-encapsulated gold nanoparticles. J. Phys. Chem. B **110**(35), 17478–17483 (2006)
74. Crespilho, F.N., et al.: A strategy for enzyme immobilization on layer-by-layer dendrimer-gold nanoparticle electrocatalytic membrane incorporating redox mediator. Electrochem. Commun. **8**(10), 1665–1670 (2006)
75. Rivas, G.A., et al.: Carbon nanotubes for electrochemical biosensing. Talanta **74**(3), 291–307 (2007)
76. Zheng, W., et al.: A glucose/O_2 biofuel cell base on nanographene platelet-modified electrodes. Electrochem. Commun. **12**(7), 869–871 (2010)
77. Pumera, M.: Graphene-based nanomaterials and their electrochemistry. Chem. Soc. Rev. **39**(11), 4146–4157 (2010)
78. Fang, B., Kim, J.H., Yu, J.S.: Colloid-imprinted carbon with superb nanostructure as an efficient cathode electrocatalyst support in proton exchange membrane fuel cell. Electrochem. Commun. **10**(4), 659–662 (2008)
79. Iijima, S.: Helical microtubules of graphitic carbon. Nature, **354**(6348), 56–58 (1991)
80. Ajayan, P.M.: Nanotubes from carbon. Chem. Rev. **99**(7), 1787–1800 (1999)
81. Chng, E.L.K., Pumera, M.: Metallic impurities are responsible for electrocatalytic behavior of carbon nanotubes towards sulfides. Chem. Asian J. **6**, 2304 (2011)
82. Liu, G., Lin, Y.: Amperometric glucose biosensor based on self-assembling glucose oxidase on carbon nanotubes. Electrochem. Commun. **8**(2), 251–256 (2006)
83. Bonanni, A., Esplandiu, M.J., del Valle, M.: Impedimetric genosensors employing COOH-modified carbon nanotube screen-printed electrodes. Biosens. Bioelectron. **24**(9), 2885–2891 (2009)
84. Sharma, M.K., et al.: Highly sensitive amperometric immunosensor for detection of Plasmodium falciparum histidine-rich protein 2 in serum of humans with malaria: comparison with a commercial kit. J. Clin. Microbiol. **46**(11), 3759 (2008)
85. Li, J., et al.: Carbon nanotube nanoelectrode array for ultrasensitive DNA detection. Nano Lett. **3**(5), 597–602 (2003)
86. Valota, A.T., et al.: Electrochemical behavior of monolayer and bilayer graphene. ACS Nano **5**(11), 8809–8815 (2011)
87. Rao, C.N.R., et al.: Some novel attributes of graphene. J. Phys. Chem. Lett. **1**(2), 572–580 (2010)
88. Kamat, P.V.: Graphene-based nanoarchitectures. Anchoring semiconductor and metal nanoparticles on a two-dimensional carbon support. J. Phys. Chem. Lett. **1**(2), 520–527 (2009)
89. Kosynkin, D.V., et al.: Longitudinal unzipping of carbon nanotubes to form graphene nanoribbons. Nature, **458**(7240), 872–876 (2009)
90. Shan, C., et al.: Direct electrochemistry of glucose oxidase and biosensing for glucose based on graphene. Anal. Chem. **81**(6), 2378–2382 (2009)
91. Kang, X., et al.: Glucose Oxidase-graphene-chitosan modified electrode for direct electrochemistry and glucose sensing. Biosens. Bioelectron. **25**(4), 901–905 (2009)
92. Goncalves, W., Lanfredi, A.J.C., Crespilho, F.N.: Development of numerical methods for signal smoothing and noise modeling in single wire-based electrochemical biosensors. J. Phys. Chem. C **115** (32), 16172–16179 (2011)

93. Baughman, R.H., Zakhidov, A.A., De Heer, W.A.: Carbon nanotubes–the route toward applications. Science **297**(5582), 787 (2002)
94. Kong, J., et al.: Nanotube molecular wires as chemical sensors. Science **287**(5453), 622 (2000)
95. Hoeben, F.J.M., et al.: Toward single-enzyme molecule electrochemistry:[NiFe]-hydrogenase protein film voltammetry at nanoelectrodes. ACS Nano **2**(12), 2497–2504 (2008)
96. Wilson, G.S., Gifford, R.: Biosensors for real-time in vivo measurements. Biosens. Bioelectron. **20**(12), 2388–2403 (2005)
97. Fishilevich, S., et al.: Surface display of redox enzymes in microbial fuel cells. J. Am. Chem. Soc. **131**(34), 12052–12053
98. Moehlenbrock, M.J., Minteer, S.D.: Extended lifetime biofuel cells. Chem. Soc. Rev. **37**(6), 1188–1196 (2008)
99. Rubenwolf, S., et al.: Strategies to extend the lifetime of bioelectrochemical enzyme electrodes for biosensing and biofuel cell applications. Appl. Microbiol. Biotechnol. **89**(5), 1315–1322 (2011)
100. Gifford, R., et al.: Protein interactions with subcutaneously implanted biosensors. Biomaterials **27**(12), 2587–2598 (2006)
101. Wisniewski, N., et al.: Decreased analyte transport through implanted membranes: differentiation of biofouling from tissue effects. J. Biomed. Mater. Res. **57**(4), 513–521 (2001)
102. Wisniewski, N., Moussy, F., Reichert, W.M.: Characterization of implantable biosensor membrane biofouling. Fresenius' J. Anal. Chem. **366**(6), 611–621 (2000)
103. Kerner, W., et al.: The function of a hydrogen peroxide-detecting electroenzymatic glucose electrode is markedly impaired in human sub-cutaneous tissue and plasma. Biosens. Bioelectron. **8**(9–10), 473–482 (1993)
104. Hahm, J., Lieber, C.M.: Direct ultrasensitive electrical detection of DNA and DNA sequence variations using nanowire nanosensors. Nano Lett. **4**(1), 51–54 (2004)
105. Vaddiraju, S., et al.: Emerging synergy between nanotechnology and implantable biosensors: a review. Biosens. Bioelectron. **25**(7), 1553–1565 (2010)
106. Kvist, P.H., et al.: Biocompatibility of an enzyme-based, electrochemical glucose sensor for short-term implantation in the subcutis. Diabetes Technol Ther. **8**(5), 546–559 (2006)

Chapter 3
Nanomaterials for Enzyme Biofuel Cells

Serge Cosnier, Alan Le Goff and Michael Holzinger

Abstract This book chapter describes the recent advances in the design of novel materials for enzymatic fuel cells. Energy conversion using biologic catalysts became a steady growing research field for supplying nomad or implantable devices due to the high specifity for the substrates and the high efficiency of redox enzymes. The constant issue, however, is the electric connection of the enzymatic redox centre to the electrode to obtain a high efficient biofuel cell. Among many advantages, nanotechnology have been offering exciting tools to achieve efficient interfacing between redox enzymes and electrical circuitry, while providing high active surfaces. We briefly introduce the principles that govern the production of electrical energy from biofuels using a biofuel cell. We focus our discussion on nanomaterials that have realized the efficient immobilization and wiring of enzymes, in particular carbon nanotubes, inorganic and polymer nanoparticles. We highlight the successfull use of these advanced materials in the engineering of enzyme electrodes and the design of novel miniaturized biofuel cell setups.

3.1 Introduction

Growing demand for energy in our modern society combined with a medium-term depletion of fossil fuels and the environmental impact of combustion of fossil energy, imply to find other modes of energy production. Among the novel sources of clean energy without greenhouse gas emissions or environmental pollution, the

---

S. Cosnier (✉) · A. Le Goff · M. Holzinger
Département de Chimie Moléculaire, UMR-5250, ICMG FR-2607, CNRS, Université Joseph Fourier, BP-53, 38041 Grenoble Cedex 9, France
e-mail: Serge.Cosnier@ujf-grenoble.fr

F. N. Crespilho (ed.), *Nanobioelectrochemistry*, DOI: 10.1007/978-3-642-29250-7_3,

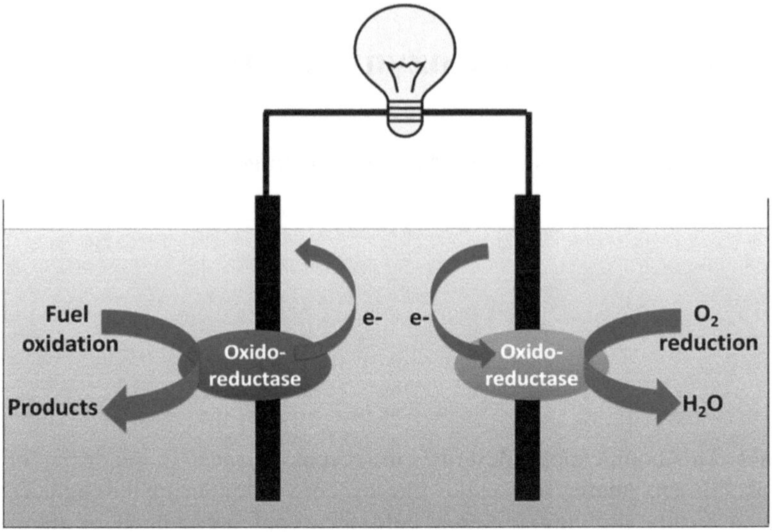

Fig. 3.1 Schematic representation of an enzymatic biofuel cell design based on the electrical connection of a laccase at the cathode and glucose oxidase at the anode

production of energy through electrochemical means such as fuel cells is a global challenge.

A subcategory within the fuel cell topic concerns biofuel cells. Such investigations occupy a prominent place in global research in transforming chemical energy into electrical energy by the bio-catalytic reaction of enzymes or living organisms. Enzymatic biofuel cell design primarily involves the use of redox enzymes for the oxidation of targeted specific fuels (sugars, alcohols or hydrogen) at the anode and the reduction of oxidants (O_2, H_2O_2) at the cathode to generate electrical power (Fig. 3.1). Taking into account that enzymes have high specific activity and are very selective, the design of enzymatic biofuel cells does not necessarily require a separation between the bioanode and the biocathode unlike the configuration of common fuel cells [1–5].

It should also be emphasized the ecological aspects inherent to biofuel cells that contrarily to fuel cells, require no metal catalysts (platinum, nickel, palladium, rhodium, iridium, etc.). Indeed, materials, fuels, and products used in the design of all biofuel cells are biodegradable. Consequently, these biofuel cells are not subjected to major economic issues related to metal catalyst. Indeed, the increasing demand for strategic metals and metal alloys by high-tech industries, aerospace or automotive industry causes a process of depletion of these materials.

The scientific challenge of these enzymatic biofuel cells is to develop devices with compatible power and size to use them as power sources for portable devices such as GPS, mobile phone, MP3 players, or mobile computers. A steady increasing interest within enzymatic biofuel cell design is dedicated to the production of electrical energy from the electro-enzymatic degradation of glucose and O_2. These two compounds are present in body fluids (blood or extracellular fluids), and therefore the main motivation for the development of biofuel cells is focused on their potential use in the human body as an energy source for implanted medical devices. In the body, these generators will be fully autonomous and can operate pacemakers, micro machines, micro-pumps, sensors, etc.

Although the first example of a biofuel cell was described in 1964, this research topic remained virtually unexplored until the late 90s. Since the early 2000s, renewed interest has been focused on biofuel cells as evidenced by the exponential increase of scientific publications devoted to this topic (5, 55 and 265, respectively in 2000, 2005, 2010). This behavior may be explained by technological barriers preventing the development of enzymatic biofuel cells where more and more of such obstacles were circumvented like the commercial availability of a wide range of purified enzymes, the design of new biomaterials for the immobilization of a high density of enzymes and their electrical connection. In particular, the technological advances in the field of biosensors in the years 1990–2000 such as the rapid development of various procedures of enzyme immobilization as well as the availability of redox mediators to establish electrical wiring between biomolecules and the electrode, opened up vast possibilities in the field of enzymatic biofuel cells.

In this context, nanomaterials like redox clays, metallic nanoparticles or nano-objects such as carbon nanotubes, have played an important role for interfacing enzymes with electronic circuitry. In particular, these nanomaterials constitute a versatile tool for the development of three-dimensional biomaterials dedicated to improve the performance of the bioelectrodes. Moreover, nanomaterials constitute a new generation of host matrices for biological macromolecules. This may confer novel multi-functionalities to biocoatings through their own specific properties (electronic conductivity, magnetism, redox properties, affinity interactions) at the nanoscale level. In particular, these nanomaterials can establish an electrical communication with enzymes via their intrinsic conductivity (like carbon nanotubes) or via an electron transport to enzymes ensured by electron hopping between immobilized redox centers. As a consequence, electrodes modified by nanomaterials have aroused widespread attention in the design of biofuel cells.

3.1.1 Principles of Biofuel Cell Functioning: Mediated or Direct Electron Transfer

The vast majority of enzyme biofuel cells is based on the electroenzymatic oxidation of glucose by glucose oxidase (GOX) and oxygen reduction by laccase, rarely, bilirubin oxidase, or even ascorbate oxidase. Usually two couples of redox mediators are involved in the functioning of the enzymatic biofuel cell. One is required to establish an electrical connection between the electrode surface and the reduced form of flavin adenine dinucleotide, the prosthetic center of GOX. The second couple, located at the cathode, allows the electron transfer from the electrode surface to the copper center of laccase where the oxygen reduction takes place (Fig. 3.2).

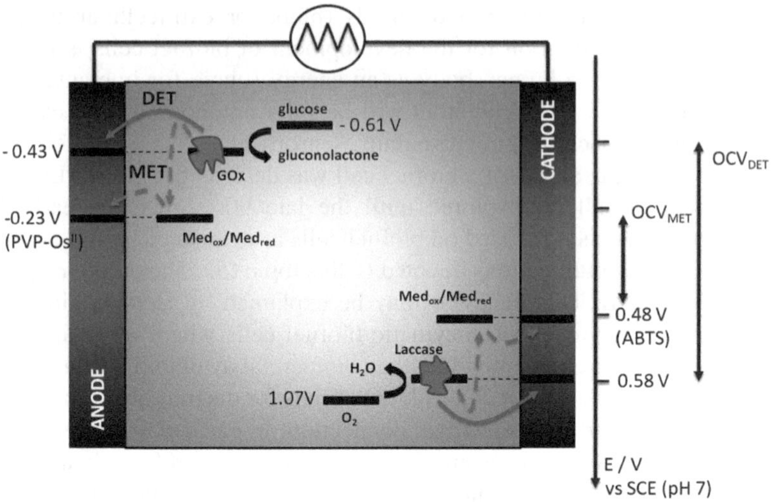

Fig. 3.2 Schematic representation of a biofuel cell design based on the electrical connection of a laccase at the cathode and glucose oxidase at the anode

In nature, electron tranfer within enzymes is realized by one or several cofactors such as NADH, PQQ, or FAD in case of glucose oxidation, or iron-sulfur clusters in hygrogenases. When artificially electron transfer to enzymes, immobilized at electrodes is required, two possible routes are mostly applied:

• Mediated electron Transfer (MET).

This implies the immobilization of a natural cofactor of the enzyme or an articial redox mediator on the electrode. As mentioned before, the redox system have to be reversible and with high electron transfer rates. The redox potential has to be as closed as possible to the redox potential of the active site of the enzyme to maximize the final OCV of the cell.

• Direct Electron Transfer (DET).

Direct electrical wiring of the enzyme to the electrode is established when the active redox center can directly be regenerated by the electrode. In this case, by considering different structures of enzymes and its location of the active site inside the protein, different strategies for their wiring are to evaluate. Indeed, direct electron tranfer becomes a challenge when the active site is deeply embedded inside the protein and cannot exchange electrons without the need of redox mediators.

The power of biofuel cells is directly related to the difference between the respective redox potentials of the electroenzymatic reactions occurring at each of the two electrodes; the bioanode for glucose oxidation and the biocathode for the reduction of oxygen. The cell voltage and hence the power thus depends on the mode of enzyme wiring and therefore, the direct electron transfer is the most attractive strategy. Indeed, redox mediators should have a redox potential close to the prosthetic site of the enzyme but should provide an efficient driving force. Consequently, biofuel cells based on mediated reactions should present a lower open circuit potential than those based on direct electron transfer.

Taking these arguments into account, two important aspects are decisive in the construction of nanostructured electrodes. Firstly, a nanostructured surface has to provides a high three-dimensional surface area that enables a high density deposition of both, the redox mediator and the enzyme. Secondly, the specific morphology of nanowires or nanoparticles have to enable intimate interactions with enzymes and finally favour the direct electrical wiring between the bulk electrode and the redox active site. An efficient direct electron transfer represents the ideal configuration for biofuel cells since no further fabrication steps are needed to introduce a redox mediator. In this case, the OCV is maximized and stability of these bioelectrodes is increased as it generally depends on the intrinsic stability of redox parnters more than of stability of enzymes.

3.1.2 Characterization of Biofuel Cell Performances

The performances of a biofuel cell can be defined by its maximum power density, its maximum current density, and the open circuit voltage. The determination of the operational stability (evolution of the biofuel cell voltage versus time for a discharge at constant current) and the storage stability with intermittent operation also provides key basic parameters on the viability of the biofuel cell. The total amount of electricity that can generate a biofuel cell can also be estimated by recording a current constant (I) supplied by the biofuel cell with time (t) for a constant cell voltage.The recording time is fixed by the stability of the tension that should not decrease more than 30 %.

Polarization curves are to evaluate for each bioelectrodes, addressing maximum current densities and open-circuit potentials. These curves allow to estimate the limiting performances of the overall biofuel cell. Furthermore, the open-circuit potential is characteristic for the presence of overvoltages, mediated or direct electron transfer rates, and the respective efficiency of electronic communication with enzymes (Fig. 3.3a).

The main performances of biofuel cells are often represented as a power/ voltage curve or current/voltage curve (Fig. 3.3b, c).

I. The different regions of the current/voltage curve are characteristic of the biofuel cell : (i) activation polarization at low current densities where the cell mainly depends on cell overvoltages and kinetics of electron transfer, (ii) ohmic part influenced by the materials, their interfaces, and the overall resistance of the circuit and (iii) polarization concentration at high current densities where the cell is limited by mass transport of fuel and oxidizing agent. The limiting values are the maximum voltage associated with a zero current and, at the other

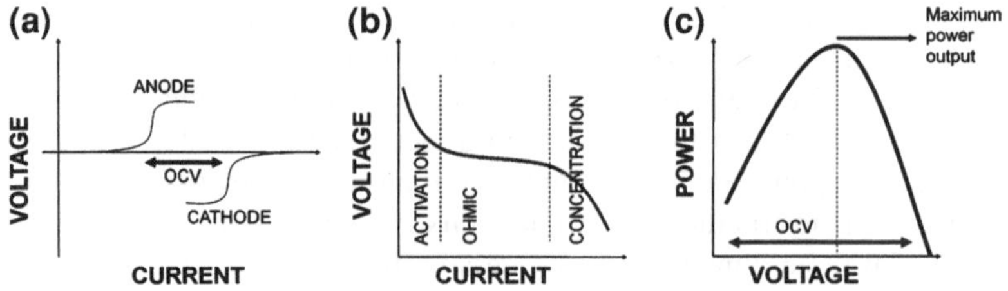

Fig. 3.3 Schematic representation of the performance of the biofuel cell: **a** polarization curves for bioelectrodes, **b** polarization curve for the biofuel cell, **c** curve of power vs. output voltage

extreme, the maximum current recorded between the electrically connected bioanode and biocathode to an external circuit represents the theoretical value without resistance. The shape of the power/voltage looks like a bell curve where its maximum represents the maximum power of the biofuel cell. In addition, the open circuit voltage is directly indicated on the x axis.

3.2 Carbon Nanotube-Based Enzymatic Biofuel Cells

Within the vast number of available nanomaterials, carbon nanotubes (CNTs) exhibit, between others, nanowire morphology, biocompatibility and excellent conductivity. These particularities are the reason why nanotubes are considered as very promising candidates in biosensor and biofuel cell devices. Nanotube based interfaces or matrices enable better approach to the active site of the enzyme by achieving electrical wiring between active sites of biomolecules and the bulk electrode. Furthermore, the possibility to add appropriate functionalities via organic functionalization enabled optimal tuning of such nanostructured electrodes by attaching specific docking sites for biomolecules or redox mediation of bio-electrochemical reactions. Furthermore, CNT modified electrodes offer a large electroactive surface together with a highly porous three-dimensional structure.

As mentioned before, two main aspects have motivated scientists to use CNTs in BFCs: their high 3D electroactive area that increase surface concentration of enzymes and others redox partners, and their ability to access the embedded active site of the enzyme in order to achieve direct electron transfer (Fig. 3.4a).

3.2.1 Carbon Nanotubes for 3D Electrodes

The exceptionnal properties of carbon nanotubes in terms of conductivity and high electroactive surface have made them an ideal material for immobilizing numerous biomolecules leading to many applications in biosensors and biofuel cells. Another advantage lies in the full range of existing functionalization methods to confer to

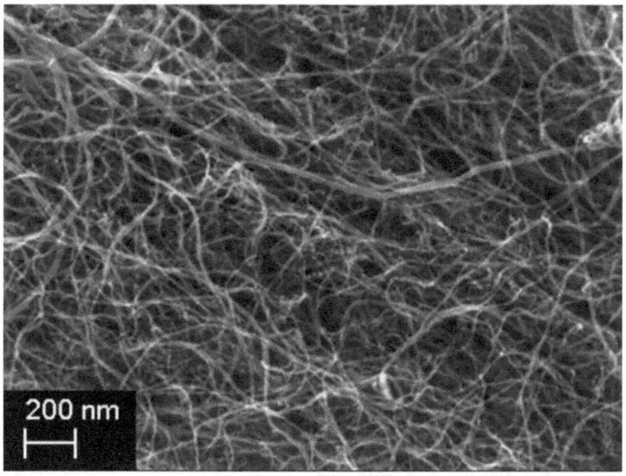

Fig. 3.4 SEM micrograph of a MWCNT electrode

carbon-based nanoparticles the ability to attach biomolecules and shuttle the electrons between the active site and the electrode. For this purpose, the combination of a carbon nanotube matrix with redox molecules, able to oxidize or to reduce efficiently the active site of enzymes, were investigated using different routes:

- Functionalization of CNTs with redox molecules

Covalent functionlization of CNTs was employed in a flexible way to modify single-walled carbon nanotubes (SWCNTs) by the corresponding redox mediator prior to their deposition on electrode. R. Bilewicz et al. reported the covalent functionalization of SWCNTs with ferrocene and ABTS [6]. Immobilized ferrocene acts as a redox bridge for the electrical wiring of GOX while at the anode and ABTS-modified SWCNTs serve for the electrical connection of laccase at the cathode.

Ferrocene-modified SWCNTs and ABTS-modified SWCNTs were deposited onto one of each electrode using a liquid-crystalline matrix-monoolein cubic phase. The GBFC delivered 100 μW cm$-^{2}$ with an OCP of 0.43 V in 20 mM glucose in quiescent solution

Taking advantage of the steady increasing techniques for CNT functionalization [7], several routes were explored to attach redox molecules onto SWCNTs. Ferrocene was also attached to MWCNTs by, amide coupling, π-stacking interactions [8], aryldiazonium reduction [8] or 1,3 dipolar cycloaddition of azomethyne ylides [9] in order to establish electrical communication between the enzyme and the electrode (Fig. 3.5).

- Carbon nanotube-doped polymers

During many years, osmium-based hydrogels have been reported to exhibit the best GBFC performances. Several groups developed polypyridyl osmium complexes bound to a polyvinylpyridine (PVP) polymer backbone. GOX and bilirubine oxidase

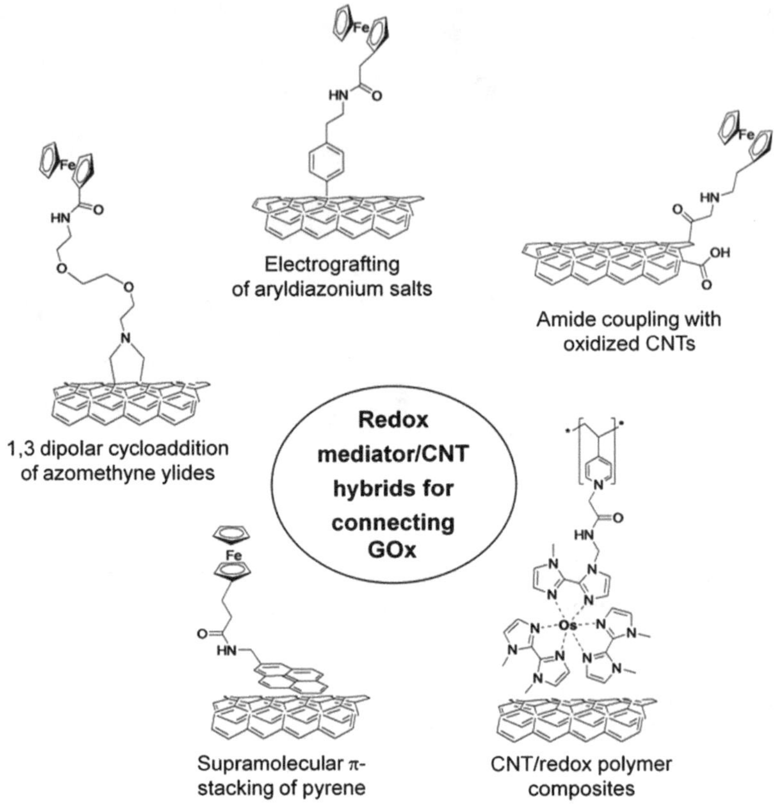

Fig. 3.5 Redox mediators combined to CNTs for indirect electrical wiring of glucose oxidase

(BOD) were further immobilized into the polymer by chemical crosslinking using poly(ethylene glycol) diglycidyl ether (PEGDGE). Different osmium complexes were succesfully synthesized using various types of n-heterocyclic ligand or electro-attractive/withdrawing groups, in order to closely approach the redox potential of the enzymes. Finally, carbon nanotube fibers made of sodium dodecyl sulfate (SDS)-dispersed carbon nanotubes injected in a poly(vinyl alcohol) matrix were used as substrate for osmium hydrogels. This configuration lead to a high performance biofuel cell exhibiting 740 μW cm^{-2} at 0.57 V, using BOD at the cathode and GOx at the anode [10].

3.2.2 Carbon Nanotubes for Direct Electron Transfer

When the issue of direct electrical communication between the active site of enzymes and the electrode can be overcame, important advantages of using DET instead of MET appeared in the fabrication and the functioning of biofuel cells. These advantages consist in fewer fabrication steps, enhanced stability over time since no molecular complexes is used, and maximization of OCV by operating the cell at the distinct redox potentials of the enzymes at both, the anode and the cathode without indirect electron transfer to a secondary redox active molecule. The biocompatibily of CNTs towards biomolecules was the starting point of the growing interest of CNTs in the engineering of bioelectronic interfaces. As several enzymes exhibit direct electrical wiring between their active site and glassy carbon electrodes, the electrical behaviour of carbon nanotubes towards direct wiring of enzymes was soon investigated. First studies on the interaction between SWCNT and redox enzymes such as GOX revealed the occurrence of DET between SWCNTs and the FAD/$FADH_2$ cofactor, which could not be obtained using bulk material electrodes [11–14].

DET was also evidenced at SWCNT electrodes for other redox proteins such as hemoglobine [15], cytochrome C [11], microperoxydases [16] or catalases[17]. The ability of CNTs to achieve DET with enzymes has lead to the design of novel biofuel cell electrodes.

Envisioned as catalysts in hydrogen biofuel cell anodes, hydrogenases (H_2ases) are enzymes that catalyse the reversible oxidation of hydrogen to protons [3, 18]

With the aim to obtain DET between H_2ases and the electrode, H_2ases have been covalently grafted to carbon nanotubes (CNTs). De Lacey and co-workers reported efficient H_2 oxidation at CNT functionalized by *Desulfovibrio Gigas* [NiFe] hydrogenases exhibiting high current densities of $\sim$0.5 mA cm^{-2} at pH 7 (T = 40 °C, 20 mV s^{-1}) [19]. Furthermore, Lojou and co-workers reported maximum catalytic current of $\sim$0.05 and $\sim$1 mA cm^{-2} (pH 7.2, T = 60 °C, 10 mV s^{-1}) at oxidized SWCNT electrodes modified with *Desulfovibrio fructosovorans* and *Aquifex aeolicus* [NiFe] hydrogenases, respectively [20, 21]

First examples of partially mediatorless-based glucose biofuel cells (GBFCs) showed DET at the laccase-modified cathode while a redox mediator still had to be used to connect an enzyme at the anode.

This first example was reported by Yan et al. [22]. The GBFC was formed using glucose-dehydrogenase connected to a SWCNT electrode via poly-(Methylene Blue) and the laccase was directly wired to the SWCNT cathode. The biofuel cell delivered 9.5 μW cm^{-2} (10 mM NAD^+, 30 mM glucose at ambiant air)

The Willner group functionalized a SWCNT anode with Nile Blue and the cofactors $NADP^+$ and NAD^+ via a phenyl boronic acid ligand [23]. Connecting the anode to a BOD cathode, one ethanol biofuel cell based on alcohol dehydrogenase delivered 23 μW cm^{-2} and one glucose biofuel cell based on glucose dehydrogenase (GDH) delivered 58 μW cm^{-2}.

In two distinct GBFC designs, NAD + dependant GDH was co-adsorbed with Methylene Green at a SWCNT anode [24] or immobilized by cross-linking using glutaraldehyde at a SWCNT anode covalently functionalized with Nile blue [25]. In both set-ups, laccase was directly wired to the cathode. These two GBFCs exhibit power output of 58 μW cm^{-2} at 0.4 V (45 mM glucose/air) and 32 μw cm^{-2} at 0.35 V (40 mM glucose/air), respectively.

Recently, the first example of a complete mediatorless glucose/O_2 biofuel was reported. The DET at a laccase/MWCNT cathode and GOx/MWCNT anode was

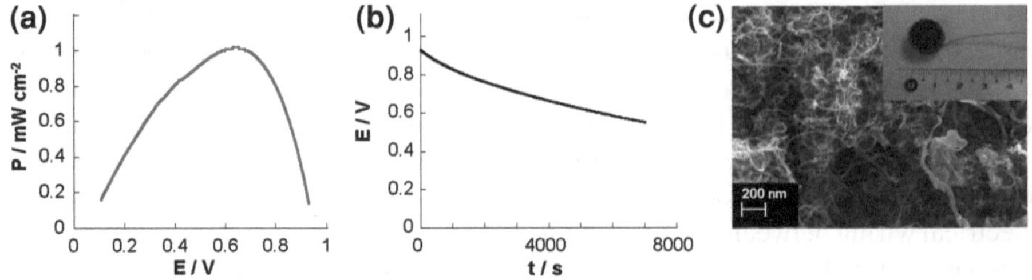

Fig. 3.6 **a** Power density vs operating voltage in 0.005 mol L^{-1} glucose solution **b** Continuous discharge under 200 μA cm^{-2} in 0.05 mol L^{-1} glucose solution **c** SEM image and (inset) photograph of the bioelectrode used for the mediatorless glucose biofuel cell

efficiently realized using a soft compression technique [26]. The biofuel cell exhibited exceptional stability over months, high power output of 1 mW cm^{-2} at low glucose concentration (5 mM glucose air saturated), and an OCV of 0.95 V in quiescient solution. Catalase, an enzyme, was employed in the GOX-based bioanode to decompose H_2O_2 (a side product of the enzymatic glucose oxidation) into O_2 and H_2O. One of the key aspect in this type of biofuel cell is the soft pressure applied to the CNT/enzyme mixture that is responsible for the efficient electrical wiring of the enzyme. A second important feature of this material is the combination of a high porosity (BET equal to 180 m^2 g^{-1}) and high conductivity (3300 S m^{-1}) that favour diffusion of substrates to the enzymes and electron mobility respectively (Fig. 3.6).

A complete mediatorless fructose/O_2 biofuel cell was also reported and based on a liquid-induced shrinkage of a free-standing MWCNT-forest film [27]. The biofuel cell delivered 1.8 mW cm^{-2} at 0.4 V in a stirred fructose solution (200 mM) using a fructose dehydrogenase anode and a laccase cathode.

3.2.3 Other Carbon-Based Nanomaterials

Beside great opportunities in the developpment of biofuel cells using carbon nanotubes, other types of carbon nanostructures have shown to be able to interface efficiently redox enzymes with electrodes.

Graphene nanoplatelets showed DET at both, bioanodes and cathodes using GOX and laccase, respectively. These connected bioelectrodes provided a maximum power output of 60 μW cm^{-2} and 0.6 V OCP [28].

Ordered mesoporous carbon also demonstrated interesting performances with a 110 μW cm^{-2} maximum power output and 1.2 V OCP [29]. Recently, the design of a miniature glucose/O_2 biofuel cell based on single-walled carbon nanohorns (SWCNHs) attached to carbon micro electrodes has been reported [30]. Electrical communication could be obtained using glucose dehydrogenase (GDH) at the anode and bilirubin oxidase (BOD) at the cathode. This setup provided a maximum power density of 140 μW cm^{-2} at 0.51 V harvested from soft drinks. The concept of mediated electron transfer by incorporation of fullerenes (C_{60}) as redox mediators inside an ordered mesoporous carbon (OMC) matrix was demonstrated by Zhou et al. This setup has been evaluated using NADH as biomimemic redox probe for different host matrices. The OMC-C_{60} configuration showed improved electron-transfer kinetics than e.g. similar nanotube designs [31].

3.3 Nanoparticle-Based Enzymatic Biofuel Cells

3.3.1 Clay Nanoparticles

Clay nanomaterials consist in layered structures built on stacked elementary nanoparticles. These nanoparticles are positively or negatively charged and are separated by interlamellar domains occupied by water molecules and exchangeable anions or cations depending on their charge. Colloidal suspensions of clay nanoparticles can easily be prepared by dispersing clays in water by stirring several hours. Clay nanoparticles are widely employed as hydrophilic additives to improve the biocompatibility of organic polymers, in particular polypyrrole and polyaniline. The latter were intensively used for the fabrication of biosensors. However, the hydrophobic character of these host films alter the three-dimensional structure of the entrapped enzymes and hence diminish the biological activity.

The unusual intercalation properties of clays were also applied to the soft and rapid immobilization of enzymes. The procedure consists in the addition of biomolecules into aqueous clay nanoparticles dispersion. The adsorption of these dispersions leads to inorganic biostructures with open frameworks.

In this context, the adsorption of an aqueous enzyme-clay mixture onto an electrode surface was widely used for biosensor fabrication [32]. Owing to the presence of microchanels within the clay matrices, a chemical crosslinking step of the protein by glutaraldehyde was often carried out in order to prevent the release of the entrapped enzymes.

The characteristics of clay coatings, such as porosity and swelling properties in aqueous solutions, induce an improvement in the activity and stability of the immobilized enzymes. The resulting biocoatings present many advantages, such as higher surface-to-volume ratio which increases susceptibility to external influences (e.g. rate of mass transport to and from an electrode) and possibilities to control the fundamental properties of the host matrices. Besides their chemical inertia and mechanical stability, the ion-exchange properties of clay nanomaterials have paved the way for large numbers of new materials of desirable properties which have useful functions for numerous electrochemical biosensor and biofuel cell applications.

In particular, Layered Double Hydroxides (LDH) constitute a promising electrode material where its structure is composed of stacked positive layers. The

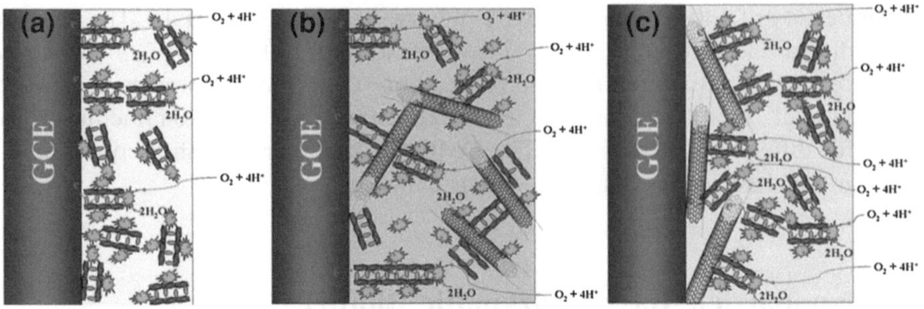

Fig. 3.7 Schematic representation of **a** LDH-ABTS laccase electrode **b** SWCNT- LDH-ABTS laccase mixed coating and **c** "two layers" configuration based on an inner SWCNT deposit modified by a LDH-ABTS laccase coating

electroneutrality of the structure is tuned by the integration of exchangeable anions accompanied with water molecules in the interlamellar domains. These electrostatic interactions were used for the intercalation of anionic organic electroactive molecules such as anthraquinone disulfonate, 2,2′-azinobis 3-ethylbenzothiazoline-6-sulfonate (ABTS), ferrocene derivatives, nitroxide or porphyrines into LDH layers conferring thus electroactive properties to the inorganic matrix. These incorporated redox mediators play the role of electron shuttles between the electrode and the active center of enzymes that is often located deep inside the protein. For instance, LDH functionalized with ABTS redox mediators was successfully applied to the immobilization and electrical wiring of peroxidase and laccase leading to electrochemical biosensors for H_2O_2 and O_2, respectively [33, 34]. In particular, dissolved oxygen was detected at a LDH-ABTS laccase electrode in a dynamic concentration range of 6×10^{-8} to 4×10^{-6} M. Laccase catalyses the four-electron reduction of oxygen directly to water by oxidizing ABTS anions. It should be noted that laccase electrodes have aroused a considerable attention as biocathode for the development of biofuel cells [35–38]. Taking the attractive potentialities of LDH-ABTS laccase electrode for oxygen reduction into account, this biomaterial was employed to develop biocathodes of biofuel cell. However, although the electron transfer within redox LDH was described as an electron hopping mechanism, one of the limitations of enzyme-clay electrodes lies in the non-conductive nature of these clay nanoparticles.

In order to improve the conductivity of the clay nanomaterials, an original approach consists in the combination of LDH nanoparticles and SWCNT. The conductive nature of SWCNT should improve the charge transport within the clay-enzyme coating and hence the electrical communication with entrapped laccase molecules. The intimate association of these nanoparticles was attempted by electrostatic interactions. For this purpose, SWCNTs were chemically oxidized for generating hydroxyls and carboxylic groups on the nanotube sidewall. This functionalization provides negatively charged SWCNT and hence facilitates their dispersion in aqueous solutions. These dispersed nanotubes can thus interact with positively charged LDH nanoparticles. Nanotubes were combined with LDH

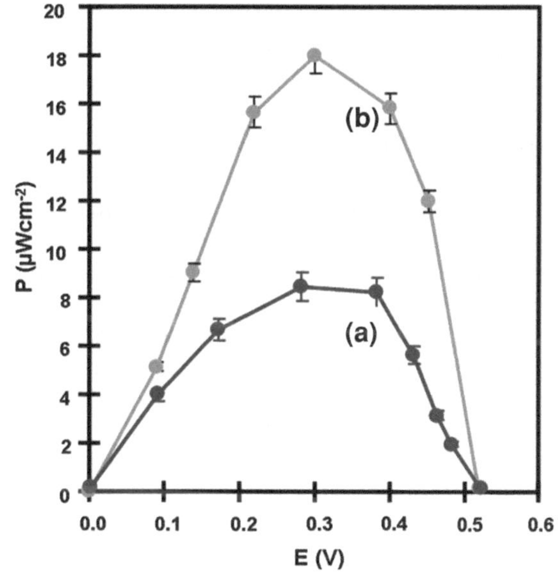

Fig. 3.8 Power density of laccase-glucose oxidase biofuel cell as a function of cell potential in air-saturated 0.1 M phosphate buffer (pH 6.0) containing 5 mM glucose for biofuel cell based on **a** LDH-ABTS laccase biocathode and **b** "two layers" configuration (SWCNT deposit modified by a LDH-ABTS laccase coating) biocathode

nanoparticles associated with ABTS following different procedures leading two different configurations. The latter correspond to a mixed deposit obtained by mixing all components in water or a « two layers » coating by creating first a SWCNT deposit and then adsorbing LDH and laccase (Fig. 3.7).

The electrocatalytic property of the different biocathode configurations for O_2 reduction was compared in term of maximum current density at 0.35 V that corresponds to the plateau of the electrocatalytic wave. In addition, the influence of the nanotube percentage within the hybrid coating on the catalytic effect was also investigated. It appears that the « two layers » design is the optimal configuration with 43 % SWCNT loading. The latter corresponds to a nanotube deposit of 215 $\mu g\ cm^{-2}$. It should be noted that the maximum current density (77.6 $\mu A\ cm^{-2}$) was markedly stronger than that (28 $\mu A\ cm^{-2}$) recorded at a LDH-ABTS laccase electrode without SWCNTs that highlights the beneficial effect of nanotubes on the enzyme wiring and electron transport processes (Fig. 3.8).

A membrane-less glucose/air biofuel cell was built by combining the biocathode based on the "two layers" configuration with a bioanode composed of glucose oxidase wired by ferrocene. The bioanode consists in a compact graphite disc (diameter 1.33 cm) prepared by mechanical compression of a mixture of graphite particles, glucose oxidase and ferrocene at 10 000 kg cm^{-2}. Figure 3.7 shows the performance of the resulting biofuel cell in air-saturated 0.1 M phosphate buffer (pH 6.0) containing 5 mM glucose. The maximum power output of the biofuel cell was 18 $\mu W\ cm^{-2}$ at 0.3 V while the open circuit voltage (OCV) reached 510 mV. In order to corroborate the beneficial role played by SWCNT within the LDH coating, a biofuel cell composed of an identical bioanode associated to a LDH-ABTS laccase electrode without SWCNTs as biocathode, was prepared and evaluated. As expected, a similar OCV was registered and the maximum power density was markedly lower, namely 8.3 μW cm^{-2}.

The redox clay nanoparticles were also incorporated in polypyrrole films and used as redox mediator for the electrical connection of entrapped enzymes. Thus, LDH-ABTS and LDH-$Fe(CN)_6$ have been synthesized and applied to the electrical wiring of laccase and glucose oxidase, respectively. For this purpose, redox LDH and enzymes were co-immobilized by entrapment in electrochemically generated polypyrrole films. This one-step method consists in the application of an appropriate potential (0.8 V vs. SCE) to a tubular porous carbon electrode soaked in an aqueous solution containing enzyme, pyrrole derivative and dispersed LDH nanoparticles [39]. Biomolecules and inorganic nanoparticles present in the immediate vicinity of the electrode surface are thus physically incorporated inside the growing network of the polymer. The polypyrrole/LDH-ABTS/laccase electrode allows the electro-enzymatic reduction of O_2, whereas the bioanode, polypyrrole/LDH-$Fe^{III}(CN)_6$/ glucose oxidase was used for the electro-enzymatic oxidation of glucose. The resulting biofuel cell was formed by two compartments separated by a Nafion membrane. This allowed the use of an optimum pH for each enzymatic reaction, namely pH 3 for laccase and pH 7 for glucose oxidase. The biofuel cell exhibited a maximum power density of 45 μW cm^{-2} at 0.2 V, the OCV being 0.37 V.

3.3.2 Metal Nanoparticles

Without taking into account that metal nanoparticles are used as electrocatalysts, metal nanoparticles represented a flexible way to immobilize enzymes at electrodes while sometimes achieving DET. Among other metals, gold nanoparticles were deeply investigated in enzymatic biosensors [40] thanks to the easy introduction of functional groups to their surface via modified thiol groups and the establishement of DET with redox enzymes such as HRP [41] or GOX [42]. However, only few examples of biofuel cells based on gold nanoparticles have been reported. Gold nanoparticle-doped polyaniline films were used to connect electrically GDH at the bioanode and BOD at the biocathode [43]. The biofuel cell delivered 32 μW cm^{-2} power output and 0.5 V OCV. DET was also evidenced at gold nanoparticle-modifed electrodes in a fructose/O_2 biofuel cell based on a fructose dehydrogenase anode and a BOD cathode, exhibiting 0.66 mW cm^{-2} at 360 mV and 0.8 V OCP [44]. An ethanol/O_2 biofuel took advantage of a gold nanoparticle sol–gel matrix based on chitosan and partially sulfonated (3-mercaptopropyl)-trimethoxysilane sol–gel [45]. The NAD^+-dependent alcohol dehydrogenase was connected via MET using Meldola's Blue while the laccase was connected via DET. The resulting biofuel cell exhibited an open-circuit voltage of 860 mV and a maximum power density of 1.56mWcm^{-2} at 550 mV.

Another approach for designing biofuel cells is to combine the bioelectrocatalytic properties of enzymes and light-harvesting nanomaterials such as metal oxide nanoparticles. Seeking for an efficient photoelectrolytic hydrogen production cell, hydrogenases were coupled to dye-sensitized titanium oxide nanoparticles. Two examples of photoelectrochemical biofuel cells were reported that underlined promising applications. Moore et al. designed a two-compartment photoelectrochemical biofuel cell [46]. One compartiment realized the photocurrent generation by Zinc porphyrin-sensitized nanostructured TiO_2 electrodes, while the other compartment produced hydrogen at pyrolytic graphite edges and carbon felt electrodes modified with FeFe hydrogenases from *Clostridium acetobutylicum.* This cell obtained production rates of 23.4 nmol H_2 min^{-1}. E. Reisner et al. performed the efficient wiring of [NiFeSe]-hydrogenase from *Desulfomicrobium baculatum* directly onto TiO_2 nanoparticles and generated hydrogen from sunlight at high turnover rates of 50 mol H_2 s^{-1} mol^{-1} (total hydrogenase) under visible light [47].

3.3.3 Other Nanomaterials

Eventhough other types of nanomaterials have interesting DET properties with enzymes used for biosensing applications, the efficicency of interfacial electron transfer and low conductivity still hamper their developpment in the field of biofuel cells. However, among other nanomaterials investigated for BFCs, another interesting approach is the design of polymer nanowires which are able to entrap enzymes and redox mediators, and provide a high specific surface. Despite the fact that this area is still in its infancy compared to carbon-nanotube BFCs, polymer nanoparticles have attractive properties arising from their ability to be easily functionalized and to entrap enzymes. Kim et al. reported a glucose/O_2 biofuel cell integrated in a fluidic system based on a hydroquinone sulphonate/GOX anode and ABTS/laccase cathode [48]. Polypyrrole electrodes were designed by electropolymerization within the pores of anodized aluminium oxide as template.

An original use of polymer nanowires as proton conductor was investigated in designing nanobiofuel cells. A nanobiofuel cell, made of a single Nafion/poly (vinyl pyrrolidone) nanowire (200 to 800 nm thick) connected a laccase cathode and a GOX/CNT anode which delivered from 0.8 to 3 μW [49].

3.4 Conclusion

Biofuel cells attract more and more attention as green and non polluant energy source for, in general, mobile and implantable devices. Within this research topic nanostructured materials prevail due to their higher efficiency, energy yields, and the possiblity to construct miniaturized devices. At present, carbon nanotubes seem to be the most appropriate electrical host matrix in biofuel cells due to their biocompability, high conductivity, high specific surface and ability to electrically connect many redox enzymes. Nevertheless, other carbon or metal based nanostructures also show particular interesting suitabilities and represent promising alternatives, especially in microfluidics and photovoltaics approaches.

References

1. Barton, S.C., Gallaway, J., Atanassov, P.: Enzymatic biofuel cells for implantable and microscale devices. Chem. Rev. **104**(10), 4867–4886 (2004). doi:10.1021/cr020719k
2. Atanassov, P., Apblett, C., Banta, S., Brozik, S., Barton, S.C., Cooney, M., Liaw, B.Y., Sanjeev Mukerjee, Minteer, S.D.: Enzymatic biofuel cells. The Electrochemical Soc Interface **16**(2), 28–31 (2007).
3. Cracknell, J.A., Vincent, K.A., Armstrong, F.A.: Enzymes as working or inspirational electrocatalysts for fuel cells and electrolysis. Chem. Rev. **108**(7), 2439–2461 (2008). doi:10.1021/cr0680639
4. Zayats, M., Willner, B., Willner, I.: Design of amperometric biosensors and biofuel cells by the reconstitution of electrically contacted enzyme electrodes. Electroanalysis **20**(6), 583–601 (2008). doi:10.1002/elan.200704128
5. Willner, I., Yan, Y.-M., Willner, B., Tel-Vered, R.: Integrated enzyme-based biofuel cells–a review. Fuel Cells **9**(1), 7–24 (2009). doi:10.1002/fuce.200800115
6. Nazaruk, E., Sadowska, K., Biernat, J., Rogalski, J., Ginalska, G., Bilewicz, R.: Enzymatic electrodes nanostructured with functionalized carbon nanotubes for biofuel cell applications. Anal. Bioanal. Chem. 398(4), 1651–1660 (2010). doi:10.1007/s00216-010-4012-1
7. Singh, P., Campidelli, S., Giordani, S., Bonifazi, D., Bianco, A., Prato, M.: Organic functionalisation and characterisation of single-walled carbon nanotubes Chem. Soc. Rev. 38, 2214–2230 (2009). doi:10.1039/B518111A
8. Le Goff, A., Moggia, F., Debou, N., Jegou, P., Artero, V., Fontecave, M., Jousselme, B., Palacin, S.: Facile and tunable functionalization of carbon nanotube electrodes with ferrocene by covalent coupling and π-stacking interactions and their relevance to glucose biosensing. J. Electroanal. Chem. 641(1–2), 57–63 (2010). doi:10.1016/j.jelechem.2010.01.014
9. Callegari, A., Cosnier, S., Marcaccio, M., Paolucci, D., Paolucci, F., Georgakilas, V., Tagmatarchis, N., Vázquez, E., Prato, M.: Functionalised single wall carbon nanotubes/ polypyrrole composites for the preparation of amperometric glucose biosensors. J. Mater. Chem. **14**, 807–810 (2004). doi:10.1039/b316806a
10. Gao, F., Viry, L., Maugey, M., Poulin, P., Mano, N.: Engineering hybrid nanotube wires for high-power biofuel cells. Nat Commun **1**(1), 1–7 (2010)
11. Guiseppi-Elie, A., Lei, C.H., Baughman, R.H.: Direct electron transfer of glucose oxidase on carbon nanotubes. Nanotechnology **13**(5), 559–564 (2002). doi:10.1088/0957-4484/13/5/303
12. Liu, J., Chou, A., Rahmat, W., Paddon-Row, M.N., Gooding, J.J.: Achieving direct electrical connection to glucose oxidase using aligned single walled carbon nanotube arrays. Electroanalysis **17**(1), 38–46 (2005). doi:10.1002/elan.200403116
13. Patolsky, F., Weizmann, Y., Willner, I.: Long-range electrical contacting of redox enzymes by SWCNT connectors. Angew. Chem., Int. Ed. **43**(14), 2113–2117 (2004). doi:10.1002/ anie.200353275
14. Vaze, A., Hussain, N., Tang, C., Leech, D., Rusling, J.: Biocatalytic anode for glucose oxidation utilizing carbon nanotubes for direct electron transfer with glucose oxidase Electrochem. Comm. **11**(10), 2004–2007 (2009). doi:10.1016/j.elecom.2009.08.039
15. Cai, C., Chen, J.: Direct electron transfer and bioelectrocatalysis of hemoglobin at a carbon nanotube electrode Anal. Biochem. **325**(2), 285–292 (2004). doi:10.1016/j.ab.2003.10.040
16. Gooding, J.J., Wibowo, R., Liu, J., Yang, W., Losic, D., Orbons, S., Mearns, F.J., Shapter, J.G., Hibbert, D.B.: Protein electrochemistry using aligned carbon nanotube arrays. J. Am. Chem. Soc. **125**(30), 9006–9007 (2003)
17. Wang, L., Wang, J., Zhou, F.: Direct electrochemistry of catalase at a gold electrode modified with single-wall carbon nanotubes. Electroanalysis **16**(8), 627–632 (2004). doi:10.1002/ elan.200302849
18. Wait, A.F., Parkin, A., Morley, G.M., dos Santos, L., Armstrong, F.A.: Characteristics of enzyme-based hydrogen fuel cells using an oxygen-tolerant hydrogenase as the anodic catalyst. J. Phys. Chem. C **114**(27), 12003–12009 (2010). doi:10.1021/jp102616m

19. Alonso-Lomillo, M.A., Ruediger, O., Maroto-Valiente, A., Velez, M., Rodriguez-Ramos, I., Munoz, F.J., Fernandez, V.M., De Lacey, A.L.: Hydrogenase-coated carbon nanotubes for efficient H2 oxidation. Nano Lett. **7**(6), 1603–1608 (2007). doi:10.1021/nl070519u
20. Lojou, E., Luo, X., Brugna, M., Candoni, N., Dementin, S., Giudici-Orticoni, M.T.: Biocatalysts for fuel cells: efficient hydrogenase orientation for H2 oxidation at electrodes modified with carbon nanotubes. J. Biol. Inorg. Chem. **13**(7), 1157 (2008). doi:10.1007/s00775-008-0401-8
21. Luo, X., Brugna, M., Tron-Infossi, P., Giudici-Orticoni, M.T., Lojou, É.: Immobilization of the hyperthermophilic hydrogenase from Aquifex aeolicus bacterium onto gold and carbon nanotube electrodes for efficient H2 oxidation. J. Biol. Inorg. Chem. **14**(8), 1275–1288 (2009). doi:10.1007/s00775-009-0572-y
22. Yan, Y., Zheng, W., L. Su, L.M.: Carbon-nanotube-based glucose/O2 biofuel cells. Adv. Mater. **18**(19), 2639–2643 (2006). doi:10.1002/adma.200600028
23. Yan, Y.-M., Yehezkeli, O., Willner, I.: Integrated, electrically contacted NAD(P) + -dependent enzyme–carbon nanotube electrodes for biosensors and biofuel cell applications. Chem. Eur. J. **13**(36), 10168–10175 (2007). doi:10.1002/chem.200700806
24. Li, X., Zhou, H., Yu, P., Su, L., Ohsaka, T., Mao, L.: A miniature glucose/O2 biofuel cell with single-walled carbon nanotubes-modified carbon fiber microelectrodes as the substrate. Electrochem. Comm. **10**(6), 851–854 (2008). doi:10.1016/j.elecom.2008.03.019
25. Saleh, F.S., Mao, L., Ohsaka, T.: Development of a dehydrogenase-based glucose anode using a molecular assembly composed of nile blue and functionalized SWCNTs and its applications to a glucose sensor and glucose/O2 biofuel cell Sens. Actuators B **152**(1), 130–135 (2011). doi:10.1016/j.snb.2010.07.054
26. Zebda, A., Gondran, C., Le Goff, A., Holzinger, M., Cinquin, P., Cosnier, S.: Mediatorless high-power glucose biofuel cells based on compressed carbon nanotube-enzyme electrodes. Nature communications (doi:10.1038/ncomms1365) (2011, in press). doi:10.1038/ncomms1365
27. Miyake, T., Yoshino, S., Yamada, T., Hata, K., Nishizawa, M.: Self-regulating enzyme—nanotube ensemble films and their application as flexible electrodes for biofuel cells. J. Am. Chem. Soc. (doi: 10.1021/ja111517e) (2011, in press). doi:10.1021/ja111517e
28. Zheng, W., Zhao, H.Y., Zhang, J.X., Zhou, H.M., Xu, X.X., Zheng, Y.F., Wang, Y.B., Cheng, Y., Jang, B.Z.: A glucose/O2 biofuel cell base on nanographene platelet-modified electrodes. Electrochem. Comm. **12**(7), 869–871 (2010). doi:10.1016/j.elecom.2010.04.006
29. Guo, C.X., Hu, F.P., Lou, X.W., Li, C.M.: High-performance biofuel cell made with hydrophilic ordered mesoporous carbon as electrode material. J. Power Source **195**(13), 4090–4097 (2010). doi:10.1016/j.jpowsour.2010.01.071
30. Wen, D., Xu, X., Dong, S.: A single-walled carbon nanohorn-based miniature glucose/air biofuel cell for harvesting energy from soft drinks. Energy & Environ. Sci. **4**(4), 1358–1363 (2011). doi:10.1039/C0EE00080A
31. Zhou, M., Guo, J., Guo, L.-p., Bai, J.: Electrochemical sensing platform based on the highly ordered mesoporous carbon-fullerene system. Anal. Chem. **80**(12), 4642–4650 (2008). doi:10.1021/ac702496k
32. Mousty, C.: Biosensing applications of clay-modified electrodes: a review Anal. Bioanal. Chem. **396**(1), 315–325 (2010). doi:10.1007/s00216-009-3274-y
33. Shan, D., Cosnier, S., Mousty, C.: HRP wiring by redox active layered double hydroxides: application to the mediated H2O2 detection. Anal. Lett. **36**(5), 909–922 (2003). doi:10.1081/AL-120019252
34. Mousty, C., Vieille, L., Cosnier, S.: Laccase immobilization in redox active layered double hydroxides: A reagentless amperometric biosensor Biosens. Bioelectron. **22**(8), 1733–1738 (2007). doi:10.1016/j.bios.2006.08.020
35. Brunel, L., Denele, J., Servat, K., Kokoh, K.B., Jolivalt, C., Innocent, C., Cretin, M., Rolland, M., Tingry, S.: Oxygen transport through laccase biocathodes for a membrane-less glucose/O2 biofuel cell. Electrochem. Commun. **9**(2), 331–336 (2007). doi:10.1016/j.elecom.2006.09.021

36. Deng, L., Shang, L., Wang, Y., Wang, T., Chen, H., Dong, S.: Multilayer structured carbon nanotubes/poly-L-lysine/laccase composite cathode for glucose/O2 biofuel cell electrochem. Comm. **10**(7), 1012–1015 (2008). doi:10.1016/j.elecom.2008.05.001
37. Boland, S., Jenkins, P., Kavanagh, P., Leech, D.: Biocatalytic fuel cells: A comparison of surface pre-treatments for anchoring biocatalytic redox films on electrode surfaces. J. Electroanal. Chem. **626**(1–2), 111–115 (2009). doi:10.1016/j.jelechem.2008.11.010
38. Tan, Y., Deng, W., Ge, B., Xie, Q., Huang, J., Yao, S.: Biofuel cell and phenolic biosensor based on acid-resistant laccase–glutaraldehyde functionalized chitosan–multiwalled carbon nanotubes nanocomposite film. Biosens. Bioelectron. **24**(7), 2225–2231 (2009). doi:10.1016/j.bios.2008.11.026
39. Cosnier, S.: Recent advances in biological sensors based on electrogenerated polymers: a review. Anal. Lett. **40**(7), 1260–1279 (2007). doi:10.1080/00032710701326643
40. Pingarrón, J.M., Yáñez-Sedeño, P., González-Cortés, A.: Gold nanoparticle-based electrochemical biosensors. Electrochim. Acta. **53**(19), 5848–5866 (2008). doi:10.1016/j.electacta.2008.03.005
41. Yi, X., Huang-Xian, J., Hong-Yuan, C.: Direct electrochemistry of horseradish peroxidase immobilized on a colloid/cysteamine-modified gold Electrode anal. Biochem. **278**(1), 22–28 (2000). doi:10.1006/abio.1999.4360
42. Liu, S., Ju, H.: Reagentless glucose biosensor based on direct electron transfer of glucose oxidase immobilized on colloidal gold modified carbon paste electrode Biosens. & Bioelectron. **19**(3), 177–183 (2003). doi:10.1016/S0956-5663(03)00172-6
43. Yehezkeli, O., Tel-Vered, R., Raichlin, S., Willner, I.: Nano-engineered flavin-dependent glucose dehydrogenase/gold nanoparticle-modified electrodes for glucose sensing and biofuel cell applications. ACS Nano **5**(3), 2385–2391 (2011). doi:10.1021/nn200313t
44. Murata, K., Kajiya, K., Nakamura, N., Ohno, H.: Direct electrochemistry of bilirubin oxidase on three-dimensional gold nanoparticle electrodes and its application in a biofuel cell. Energy Environ. Sci. **2**, 1280–1285 (2009). doi:10.1039/B912915D
45. Deng, L., Shang, L., Wen, D., Zhai, J., Dong, S.: A membraneless biofuel cell powered by ethanol and alcoholic beverage biosen. Bioelectron. **26**(1), 70–73 (2010). doi:10.1016/j.bios.2010.05.007
46. Reisner, E., Powell, D.J., Cavazza, C., Fontecilla-Camps, J.C., Armstrong, F.A.: Visible light-driven H2 production by hydrogenases attached to dye-sensitized TiO2 nanoparticles. J. Am. Chem. Soc. **131**(51), 18457–18466 (2009). doi:10.1021/ja907923r
47. Hambourger, M., Gervaldo, M., Svedruzic, D., King, P.W., Gust, D., Ghirardi, M., Moore, A.L., Moore, T.A.: [FeFe]-hydrogenase-catalyzed H2 production in a photoelectrochemical biofuel cell. J. Am. Chem. Soc. **130**(6), 2015–2022 (2008). doi:10.1021/ja077691k
48. Kim, J., Kim, S.I., Yoo, K.-H.: Polypyrrole nanowire-based enzymatic biofuel cells. Biosens Bioelectron. **25**(2), 350–355 (2009). doi:10.1016/j.bios.2009.07.020
49. Pan, C., Fang, Y., Wu, H., Ahmad, M., Luo, Z., Li, Q., Xie, J., Yan, X., Wu, L., Wang, Z.L., Zhu, J.: Generating electricity from biofluid with a nanowire-based biofuel cell for self-powered nanodevices. Adv. Mat. **22**(47), 5388–5392 (2010). doi:10.1002/adma.201002519

Chapter 4
Biosensors Based on Field-Effect Devices

José Roberto Siqueira Jr., Edson Giuliani Ramos Fernandes, Osvaldo Novais de Oliveira Jr. and Valtencir Zucolotto

Abstract This chapter brings an overview on the use of field-effect devices (FEDs) in biochemical sensors, emphasizing their advantages and specificity for biosensing, which is typical of such semiconductor-based device. Following the introductory sections on operation principles and comparison with field-effect transistors, we concentrate on different types of FEDs and their detection methods. In particular, we shall focus on ion-sensitive field-effect transistor (ISFET), electrolyte-insulator-semiconductor (EIS), light-addressable potentiometric sensor, extended-gate field-effect transistor (EGFET) and separative extended-gate field-effect transistor (SEGFET). Important contributions in the literature in biochemical sensors based on such devices are highlighted. A discussion is also provided on how the functionalization of these devices with nanostructured films can result in sensors with increased sensitivity and selectivity. Examples of modified devices containing polyelectrolytes, metallic nanoparticles, carbon nanotubes, and other compounds, used for detecting a variety of analytes, will be provided. We discuss the concepts involved in the operation principles and the particularity of different

J. R. Siqueira Jr. (✉)
Nanomaterials and Sensors Group / Institute of Exact Sciences, Natural and Education, Federal University of Triangulo Mineiro (UFTM), Uberaba-MG, 38025-180, Brazil
e-mail: jr.siqueira@fisica.uftm.edu.br

E. G. R. Fernandes · V. Zucolotto
Nanomedicine and Nanotoxicology Laboratory / Physics Institute of São Carlos, University of São Paulo, São Carlos-SP, 13566-590, Brazil
e-mail: edlaber2001@yahoo.com.br

V. Zucolotto
e-mail: zuco@ifsc.usp.br

O. N. de Oliveira Jr.
Polymer Group / Physics Institute of São Carlos, University of São Paulo, São Carlos-SP, 13566-590, Brazil
e-mail: chu@ifsc.usp.br

F. N. Crespilho (ed.), *Nanobioelectrochemistry*, DOI: 10.1007/978-3-642-29250-7_4,

FEDs. The prospects for clinical diagnosis with such biosensors and environment monitoring are also addressed. Moreover, strategies to improve sensing properties through functionalization are placed on, particularly with synergistic combination of organic and inorganic materials. For example, nanostructured films containing carbon nanotubes exhibited enhanced performance in biosensing. It is expected that this chapter may provide researchers with an alternative sensing platform to study new biochemical sensors concepts for specific applications.

4.1 Introduction

Biosensing has benefited enormously with the advance of the so-called nanobiotechnology. Over recent years, novel concepts on biosensors have appeared to cater for different applications, especially health-related systems, environmental monitoring, food control and biotechnological processes [1–4]. Among the different types of sensors and their transduction modes, bio-chemical sensors based on field-effect devices (FEDs) have deserved special attention, for they involve multidisciplinary areas, such as biochemistry, bioelectrochemistry and bioengineering, in addition to solid-state and surface physics and silicon integrated circuit technology [4–8]. The well-established silicon-based technology has been merged with nano- and biomaterials science to develop new sensors and biosensors prototypes with enhanced performance that may be applied in diverse fields [4–8]. The integration and compatibility of biomolecules with semiconductor processing and the possibility of manufacturing miniaturized sensing devices are the main advantage of such field-effect sensors [4–8].

Nanobiotechology-based biosensors have been developed with immobilization of biomolecules in miniaturized structures, which may contain hybrid materials for enhancing sensing properties [4, 9–17]. Such methods have also been applied to biosensors based on FEDs [4]. For example, carbon nanotubes (CNTs) have been used in biosensors to achieve better sensitivity and selectivity [18–22]. The key to obtain such enhanced systems is the combination of biomolecules, whose activity may be preserved for long periods of time, and nanomaterials, as CNTs, on the FEDs surface [4]. Deposition of these materials is normally done with the electrostatic layer-by-layer (LbL) technique that allows an easy control of film thickness and possible tuning of molecular architectures to yield tailored sensing units [4, 23–31].

Here we concentrate on biochemical sensors based on FEDs and their detection methods. The chapter is organized as follows. Section 4.2 describes the operation principle of a field-effect device. Different types of FEDs and their specificity are discussed in Sect. 4.3. The major results from the literature concerning biosensing, including nanotech methods, are described in Sect. 4.4, while the chapter is closed with final remarks in Sect. 4.5.

4.2 Field-Effect Devices

The semiconductor microtechnology has evolved rapidly with the advent of nanotechnology, which allowed for new sensor concepts combining chemical and biological recognition processes with silicon chip manufacturing [9, 32]. Using functional materials and silicon technologies, one may devise sensing systems, including intelligent signal processing for biochemical parameters and micro-electrodes for determining ions and metabolic products in biomedicine, food and drug analysis, environmental monitoring, defense and security purposes, including antibioterrorism and detection of biological warfare agents [5–9]. Sensors based on FEDs are suitable sensing platforms, as they offer advantages such as a small size and weight, a fast response time, robustness, integration of sensor arrays on a chip, and possible low-cost fabrication. The typical examples of FEDs are ISFETs (ion-sensitive field-effect transistors), EGFETs (extended gate field effect transistors), capacitive EIS (electrolyte-insulator-semiconductor) sensors and LAPS (light-addressable potentiometric sensors) [5–9].

FEDs are derived from metal–insulator-semiconductor capacitance or insulated-gate field-effect transistors, with the gate electrode being replaced by an electrolyte solution (test sample) and a reference electrode [5–7]. With the introduction of an additional ion- and/or charge-sensitive gate layer, biochemical FEDs are sensitive to any electrical interaction at or nearby the interface. The biochemical reactions can be detected by an ISFET, EGFET, capacitive EIS sensor or LAPS coupled with the corresponding chemical or biological recognition element. For example, changes in the chemical composition of the analyte induce changes in the FED electrical surface charge, modulating the current in the ISFET's channel, the capacitance of the EIS sensor, or the photocurrent of the LAPS. The schematic setup in Fig. 4.1 depicts the operation principle and the signal response of ISFET, EIS and LAPS structures. In common, such devices have the same transducer principle, using an electric field to create regions of excess charge in a semiconductor substrate [5–7]. For operation, the gate voltage (V_G) is applied by a reference electrode (i.e. Ag/AgCl liquid-junction electrode), which provides a stable potential in the solution, regardless of changes in dissolved species or in the pH of an analyte. The sensing information arises from the modulation of the electric field inside the insulator, resulting in a modulation of the space-charge region in the silicon at the insulator-semiconductor interface. Signal generation may come from any of the following events: pH or ion-concentration changes, ion-concentration change due to an enzymatic reaction, adsorption of charged macromolecules (e.g., polyelectrolytes, proteins, DNA), affinity binding of molecules (e.g., antigen–antibody affinity reaction, or DNA hybridization). Furthermore, changes may also arise from living biological systems, resulting from complex biochemical processes (e.g., metabolic processes of bacteria or cells, ligand-receptor interactions, action potential of nerve cells) [5–7].

For ISFET devices in particular, because of their poor isolation and the impurities penetration in the substrate in chemical environment, device encapsulation

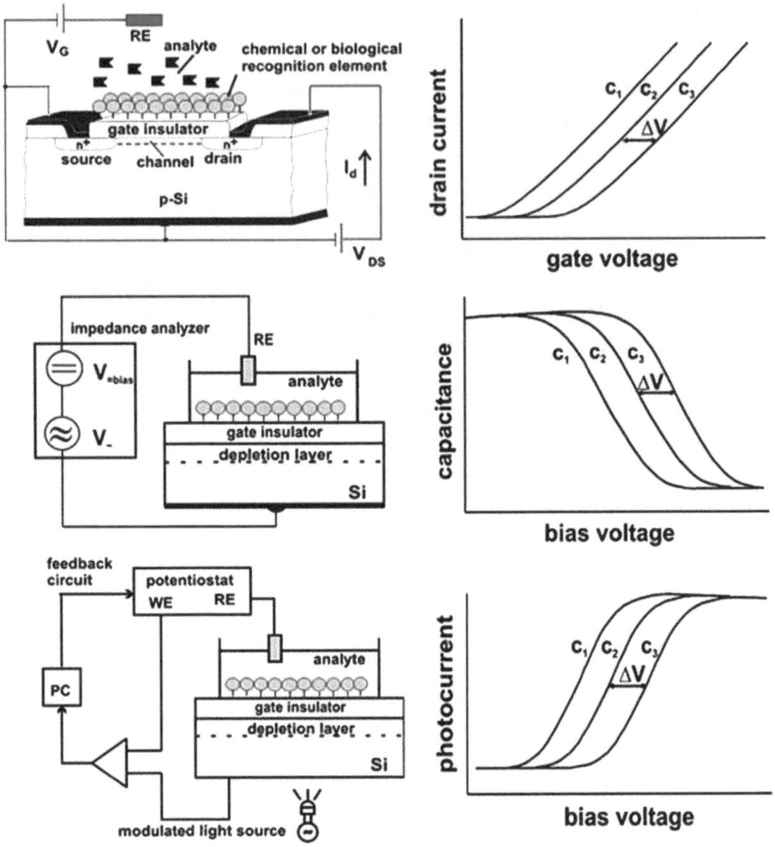

Fig. 4.1 Schematic representation of the operation principle and typical signal response of an ISFET, an EIS and a LAPS sensors. Reprinted with permission from Ref. [6]. Copyright 2012 John Wiley and Sons

was an important issue [33]. One of the alternatives for encapsulation was the use of a Si-SiO_2-Si structure, which was a complex process [34]. Another efficient strategy to isolate the FET from the chemical solution is the Extended-gate Field-effect Transistor (EGFET), shown in Fig. 4.2a, where the sensitive area is separated from the FET gate. A more advantageous configuration for sensing is a SEGFET which comprises a chemically sensitive membrane as separative extended gate (SEG) connected to the gate of a commercial field effect transistor (FET). In both cases, the circuit of the gate is closed by a reference electrode (commonly Ag/AgCl or Saturated Calomel Electrode, SCE) inserted into the chemical media (see Fig. 4.2). This configuration isolates the FET from the chemical environment, thus permitting reuse. The advantages in the latter case are: isolation from light, simple to packaging, easy fabrication, simplicity and longer stability. The difference between a SEGFET and an EGFET is the assumption of the separation of the sensitive

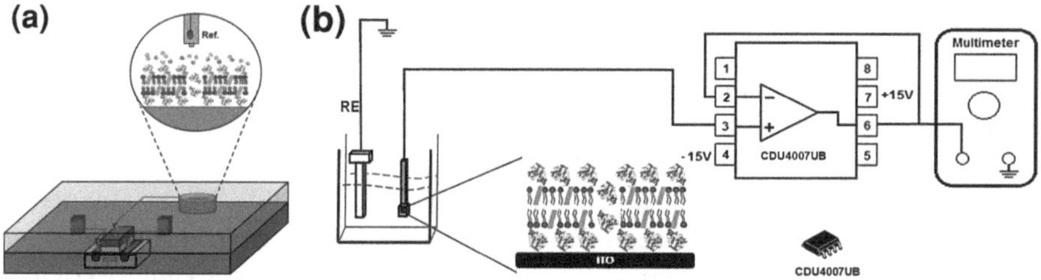

Fig. 4.2 Scheme of a biosensor based on EGFET. **a** The scheme shows a common FET device which the gate is extended from the circuit to be functionalized. The FET structure is the transducer, i.e., changes the surface potential on the extended gate into a change in the I_D. In this case, the reference electrode is floating. In **b** the scheme of a SEGFET-based biosensor (with the reference grounded) using the microchip CDU4007 as the transducer part

Table 4.1 Summary of (bio-) chemical sensors based on ISFET, EIS, LAPS, EGFET and SEGFET devices

(Bio-) chemical sensor	Ion/analyte	Sensitive membrane or (bio-) recognition element	Refs
pH sensor	H^+, OH^-	Si_3N_4, Al_2O_3, Ta_2O_5 ITO, ZnO, V_2O_5, SnO_2	[5, 35–39]
Ion sensor	K^+, Li^+, Cs^+, Ca^{2+}, Mg^{2+}, NO_3^-, SO_4^{2-}	Polymer membrane & ionophore Dendrimers, silicon nanowires	[9, 12, 20, 40–42]
Enzyme sensor	Glucose, urea, penicillin, Acetylcholine, pesticides, H_2O_2, lactate	Glucose oxidase, urease, penicillinase, acetylcholinesterase, horseradish peroxidase, organophosphorus hydrolase, phthalocyanines, L–lactic dehydrogenase	[11, 15, 16, 20, 21, 34, 43–47]

membrane from the MOSFET-based structure and absence of any microfabrication process. In the former, the sensing part is separated and connected, with a metal wire, in a commercial MOSFET (Fig. 4.2b), while in the latter the sensing part is deposited on the chip.

Examples of chemical sensors and biosensors developed using the ISFET, EGFET, SEGFET, EIS and LAPS structures as transducers are summarized in Table 4.1.

Upon applying nanotechnology methods to FEDs, one may envisage the integration of nanomaterials and biological systems into electrical devices, which is advantageous for detection of biological species, especially due to the size compatibility. In addition, the electrostatic interactions and charge transfer, typical of biological processes, may be detected by electronic nanocircuits [9, 32]. To develop nanosensor systems requires mostly infrastructure and knowledge on both silicon- and thin-film processing technologies to produce materials with control at the molecular level. In this context, the layer-by-layer (LbL) technique is promising for silicon-based sensors as this method allows a control of film architecture and thickness, in addition to the synergy between properties of distinct materials, including carbon nanotubes [48], nanoparticles [49], proteins [50], antigen–antibody pairs [51], DNA [52], and other charged macromolecules [53]. Other techniques based on chemical functionalization to modify FEDs, involving nanomaterials, such as nanowires and gold nanoparticles, have also been employed [16, 54]. The use of nanostructured LbL films in sensing has brought a number of new possibilities, mostly in terms of increased sensitivity [4, 11, 12, 15, 16, 22, 55–57].

4.3 Types of Field-Effect Devices

4.3.1 Ion-Selective Field-Effect Transistor (ISFET)

ISFET was the first miniaturized silicon-based chemical sensor, introduced by Bergveld in 1970. Since then various field-effect biochemical sensors have been developed [5–7, 33]. It derives from the insulated-gate field-effect transistor (IGFET) and from the metal–oxide–semiconductor field-effect device (MOSFET), being the most studied FED for sensing and biosensing. Its main advantage is the possible integration with signal-processing electronics on the same chip. The gate metal electrode of the IGFET is replaced by an electrolyte solution which is contacted by the reference electrode, with the insulator placed directly in an aqueous electrolyte solution [5–7], as shown in Fig. 4.1.

The operation mechanism of an ISFET is based on the flow of an electric current (I_D) from the source to the drain via the channel. Analogously to the IGFET, the channel resistance depends on the electric field perpendicular to the direction of the current, in addition to the potential difference over the gate insulator. The source-drain current (I_D) is influenced by the interface potential at the insulator/electrolyte solution. The chemical nature of the insulator interface is reflected in the measured source-drain current. According to the site-binding theory, using SiO_2 as insulator, the surface of the gate oxide contains OH-functionalities, which are in electrochemical equilibrium with ions in the sample solutions (H+ and OH-). Depending on the pH of the solution, the SiOH groups at the gate insulator surface can be protonated or deprotonated, with ensuing changes on the SiO_2 surface potential. Typical pH sensitivities measured with SiO_2-based ISFETs are 25–48 mV.pH^{-1} [5–7].

Regarding the analytical characteristics of ISFETs, the main parameters are sensitivity, selectivity, stability and drift, response time and hysteresis. Other important parameters are the linear pH range, temperature stability, light sensitivity, reproducibility and life time. These characteristics are most thoroughly studied for gate insulators made of SiO_2, Si_3N_4, Al_2O_3 and Ta_2O_5 [7, 33].

An ISFET is characterized using two basic modes, namely the constant voltage and the constant charge voltage modes. In the first, the gate voltage (V_G) and drain-source voltage (V_{DS}) are maintained constant and by measuring changes in the drain current (I_D), the pH value of a test solution can be obtained quantitatively. In the more popular constant charge voltage mode the drain current and the drain-source voltage are set at a fixed value using a feedback circuit, which causes the voltage shift. Moreover, the ISFET has to be associated with a readout-interface circuit to obtain a measuring signal. Also relevant for commercialization is the encapsulation and packaging of ISFETs. For a biochemical sensor-based ISFET, only some parts of the device need to be encapsulated, including bonding pads, bonding wires, the silicon substrate and on-chip electronics, in order to avoid damages caused by degradation of the sensing unit, liquid penetration in the electronic parts and adhesion problems [7]. Therefore, the encapsulation and packaging of ISFET should ensure electrical isolation of conducting pads, chemical isolation from the external environment, compatibility with the sensitive membrane or biocompatibility in the case of biosensors.

4.3.2 Electrolyte-Insulator-Semiconductor (EIS)

EIS is the simplest structure of a biochemical sensor based on FEDs, deriving from the capacitor metal–insulator-semiconductor (MIS) and capacitor metal–oxide–semiconductor (MOS) structures, in which the metallic gate is replaced by an electrolyte and a reference electrode [5–7]. The operation consists in applying a direct current (*dc*) polarization voltage via reference electrode to set the working point of the EIS sensor, superimposed to a small alternating current (*ac*) voltage ($\sim$10–50 mV), which is applied to the system to measure the sensor capacitance. Figure 4.1 illustrates the setup and the principle of a capacitive EIS structure.

The complete ac equivalent circuit of an EIS is complex, as it involves components such as the bulk resistance and space-charge capacitance of the semiconductor, the capacitance of the gate insulator, the interface impedance at the insulator-electrolyte interface, the double-layer capacitance, the resistance of the bulk electrolyte solution and the impedance of the reference electrode [58–60]. However, considering usual values of insulator thickness ($\sim$30–100 nm), the ionic strength of the electrolyte solution ($>10^{-4}$–10^{-5} M) and low frequencies ($<$1000 Hz), the equivalent circuit of an EIS structure can be simplified as a series connection of insulator capacitance and space-charge capacitance for the semiconductor, which is similar to the MIS capacitor [58–60]. Therefore, the capacitance of the EIS structure may be expressed in terms of the electrolyte solution/insulator interface potential (φ) as:

$$C(\phi) = \frac{C_i C_{SC}(\phi)}{C_i + C_{SC}(\phi)} \tag{4.1}$$

where C_{SC} (φ) is the space-charge capacitance, modulated by the flat-band voltage, and C_i is the insulator capacitance.

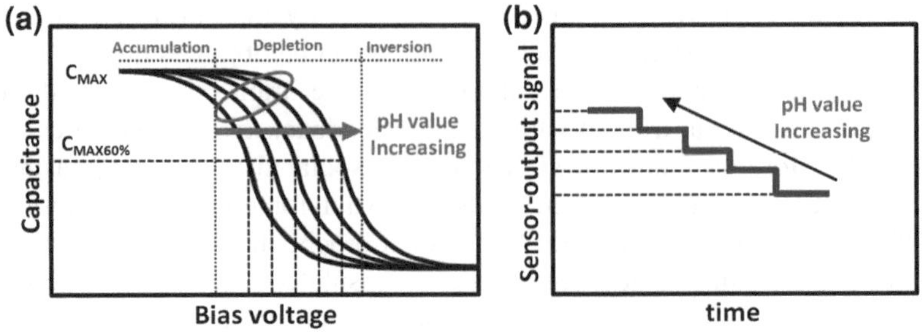

Fig. 4.3 Representation of typical C/V curves and their distinct three regions (accumulation, depletion and inversion) (**a**) and ConCap curve (**b**) at different pH values for a capacitive EIS structure

The EIS sensors are characterized using the capacitance/voltage (C/V) and constant-capacitance (ConCap) modes. Figure 4.3 depicts a typical (a) C/V curve and (b) ConCap response for a *p*-type EIS sensor at various pH values. In the C/V curve in Fig. 4.3a, three regions can be identified, namely, accumulation, depletion and inversion. For sensing, in particular, only the depletion region is considered for analysis, as the curves are shifted due to changes in the electrolyte/insulator interface potential. Thus, the C/V curves are pH or concentration-dependent for specific analytes in EIS sensors. The mechanisms for such changes were discussed in Sect. 4.2. For characterizing the chemical sensitivity of the EIS system, it is essential to keep the same conditions for the gate-insulator/semiconductor interface, in order to attribute the shifts of C/V curves entirely to the reactions at the electrolyte/insulator interface. The most important parameter to be considered is the flat-band voltage condition [7, 58–60], which can be determined by:

$$V_{fb} = E_{ref} - \phi + \chi_{sol} - \frac{W_S}{q} - \frac{Q_i + Q_{SS}}{C_i} \tag{4.2}$$

where E_{ref} is the electrode reference potential; χ_{sol} is the surface-dipole potential of the solution, and φ is the electrolyte/insulator interface potential, which depends on the activity of ions in the solution, while W_S is the semiconductor work function and Q_i and Q_{SS} are related to charges located in the insulator and the surface and interface states, respectively. The potential φ at the electrolyte/insulator interface is the only parameter which is not constant in Eq. (2.2), indicating that the sensitivity of an EIS structure, concerning the electrolyte composition, depends on changes in the flat-band voltage, which is determined by the shifts in the depletion region in C/V curves [7, 58–60].

In contrast to C/V measurements, the ConCap mode permits a dynamic investigation of sensor behaviour. This method is appropriate to set a suitable operation point and also for performing a simple, straightforward characterization of ion-sensitive layers. Furthermore, it is possible to obtain important sensing properties, including sensitivity, response time, stability, long-term and short-term drift phenomena, and hysteresis. Obtaining information with the ConCap mode is optimized if the capacitance of the working point is fixed, which corresponds to ~60–70 % of the maximum capacitance from the C/V curves. Using a feedback-control circuit to maintain this capacitance fixed, it is possible to observe shifts in the voltage owing to changes of ion-concentration at the sensor surface [7, 58–60]. A typical ConCap response is exemplified in Fig. 4.3b.

4.3.3 Light-Addressable Potentiometric Sensor (LAPS)

Differently from other field-effect sensors, the LAPS platform allows for the fabrication of a multisensory system in a same chip. The complete LAPS system consists of three units, viz., an electrochemical cell, an infrared light-emitting diode (LED) or laser as light source, and an electronic circuit to measure photocurrent [5–7]. The schematic representation of the experimental setup of the LAPS system is also shown in Fig. 4.1. The structure of the LAPS is similar to the EIS structure, since in the absence of illumination, it behaves as an EIS capacitor. Applying a *dc* bias voltage via reference electrode a depletion layer arises at the insulator/semiconductor interface. The width of this depletion layer and its capacitance vary with the insulator surface potential. The changes in capacitance in the depletion layer are detected by illuminating the LAPS chip with modulated light, which induces an *ac* photocurrent to be measured as sensor signal [61, 62].

The illumination of the semiconductor with infrared light creates electron–hole pairs, which can diffuse, recombine or be separated by an electric field. When the semiconductor is illuminated with a constant-intensity light source charge separation of photogenerated electro-hole pairs occurs in the depletion region, yielding a transient current that decays to zero as the formation of separated charges across the depletion region counteracts the tendency for a further net-charge separation [7, 61, 62]. Such behavior is somehow similar to charging a capacitor. The modulation of light intensity in a time shorter than the decay-time of the transient currents results in a modulation of the depletion-region capacitance, creating an alternated photocurrent in an external circuit. Such photocurrent appears due to the rearrangement of charge carriers in the depletion layer of the semiconductor, while the illumination is switched on and off. The amplitude of this photocurrent is the quantity to be measured. For strong depletion, the alternated photocurrent measured (I_{ph}) is determined by the electron–hole pairs, which are formed or diffused into the depletion region and by the capacitances of the illuminated area, according to Eq. 4.3:

$$I_{ph} = I_p \frac{C_i}{C_i + C_{SC}} \tag{4.3}$$

where I_p is the alternating component of the photogeneration of electron–hole pairs.

Since the capacitance of the space-charge region (C_{SC}) depends on the applied *dc* voltage, the photocurrent in the insulator/electrolyte interface is a function of the bias applied to LAPS. Such dependence is used to measure chemically sensitive surface potentials on the insulator surface [7, 61, 62]. The I/V curve is similar to the C/V curve for an EIS sensor (See Fig. 4.3), containing also the accumulation, depletion and inversion regions, but in the opposite way. Similarly to the EIS sensor, another measurement mode for LAPS is the constant-photocurrent (CC) mode, in which a feedback system controls the applied bias voltage to maintain the photocurrent constant. This mode permits to investigate dynamics, in addition to detecting solutions at different pHs or solutions at different concentrations (analyte).

4.3.4 Extended-Gate Field-Effect Transistor (EGFET)

EGFET devices were developed based on the concept introduced by Van der Spiegel et al. [63]. Figure 4.2a shows the EGFET architecture and the measurement system. EGFETs exhibit the same *I–V* operational characteristics as the MOSFET. Their pH sensitivity can be determined by measuring the drain current (I_D) as a function of the variable drain-source voltage or gate-source voltage (V_{DS} or V_{GS}, respectively) in solutions with various pHs. The measurements are made after the stabilization of the system (drift).

Based on the MOSFET equations, I_D is given in the saturation region by:

$$I_D = \frac{1}{2}\alpha(V_{GS} - V_T)^2 \tag{4.4}$$

And in the linear (Ohmic) region by:

$$I_D = \alpha[(V_{GS} - V_T)V_{DS} - \frac{1}{2}V_{DS}^2] \tag{4.5}$$

where α is a conduction parameter (geometric parameter), V_{DS} is the drain-source voltage, and V_T is the threshold voltage, defined as the minimum voltage required to make the transistor ON, which depends on the pH value [44]. Using the site-binding model proposed by Yate et al., in 1974, [64] and the Stern model for the double layer [65], it is possible to explain the pH dependence for the surface potential (Ψ_0), which depends on the membrane material and on the electrolyte pH, according to [33, 36]:

$$-2.303\Delta\text{pH} = \frac{q\Psi_0}{kT} + \sinh^{-1}\left(\frac{q\Psi_0}{kT}\frac{1}{\beta}\right) \tag{4.6}$$

Where $\Delta\text{pH} = \text{pH} - \text{pH}_{\text{pzc}}$, and pH_{pzc} corresponds to the pH at the point of zero charge on the surface, *k* is Boltzmann's constant, *q* is the elementary charge, *T* is the temperature of the system, and *B* is a dimensionless parameter that determines the relation between the pH and Ψ_0, which reflects the chemical sensitivity of the gate insulator (the parameter depends on the density of surface hydroxyl groups and the surface reactivity expressed by K_a and K_b). B is given by [66]:

$$\beta = \frac{2q^2 N_S (K_a/K_b)^{1/2}}{KTC_{DL}} \tag{4.7}$$

Where N_S is the site density, K_a and K_b are equilibrium constants, and C_{DL} is a simple capacitance (derived from the Gouy–Chapman–Stern) model.

Finally, V_T can be expressed as:

$$V_T = E_{Ref} + \chi_{sol} + V_{TM} - \frac{W_M}{q} - \Psi_0 \tag{4.8}$$

where E_{Ref} is the electrode reference potential, χ_{sol} is the surface-dipole of the electrolyte, V_{TM} is the threshold voltage of the MOSFET, and W_M is the work function of the metal gate relative to vacuum [33].

Different oxide substrates have been used as pH-sensitive membranes. However, high impedance materials used as sensitive membranes in ISFETs (SiO_2, Al_2O_3, Ta_2O_5, etc.) are not suitable for EGFET structures. Instead, low-resistance materials have been used which display Nernstian sensitivity, including indium tin oxide (ITO) [67, 68], SnO_2 [69], V_2O_5 xerogel [38, 70], ZnO [37], Vanadium/tungsten mixed oxide (V_2O_5/WO_3) [71], TiO_2:Ru [72], to name a few. Since the fabrication of such oxides is relatively expensive and involve sputtering or chemical evaporation, sol–gel methodologies have also been used [70, 73].

Organic semiconductors have been successfully applied in EGFETs because of the easy fabrication and relative low cost. In addition, organic materials may be more appropriate in nanostructured platforms for biological materials immobilization. The choice of EGFET-based biosensing is motivated by the finding that several enzymes exhibit a local pH change upon reacting with specific analytes [74–79]. Ishige et al. developed an EGFET-based biosensor using gold electrodes modified with ferrocenyl-alkanethiol and cholesterol dehydrogenase for detecting cholesterol in a concentration range from 33 to 233 $mg.dL^{-1}$ with sensitivity of 57 $mV.pH^{-1}$) [77].

4.3.5 Separative Extended-Gate Field-Effect Transistor (SEGFET)

SEGFET is a type of EGFET in which a metal wire connects the sensing film membrane to the gate of a commercial MOSFET (Fig. 4.2b), separating the FET device from the chemical environment [80, 81]. The gate-source voltage (V_{GS}) of the MOSFET is replaced by a voltage in the reference electrode (V_{Ref}) [44]. Analogously to the EGFET devices, the drain–source current can be modulated by the proton concentration on the membrane surface, upon changing the electrolyte pH. Based on Eq. 4.1, $I_D^{1/2}$ varies with pH and may present a linear pH response.

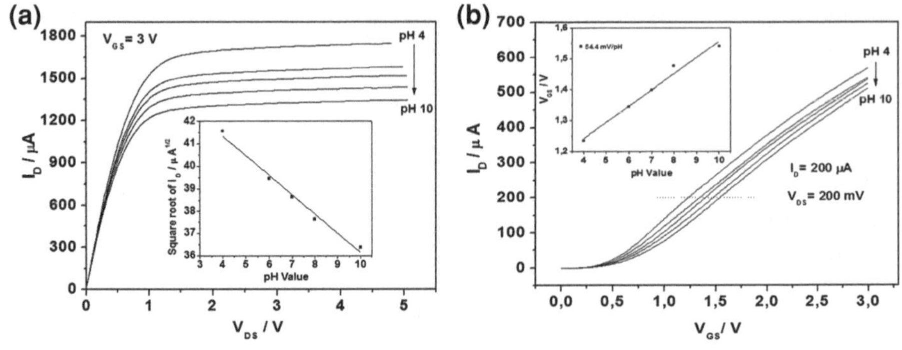

Fig. 4.4 $I_D - V_{GS}$ characteristics of the gate sensitive membrane for SEGFET (for a constant V_{GS} value). Inset: The pH dependence of the square root of I_D for the gate sensitive membrane (**a**). And $I_D - V_{GS}$ characteristics of the gate sensitive membrane for SEGFET, calculated when I_D was fixed in 200 μA. Inset: The sensitivity of the film (**b**)

Characteristic curves for separative extended gate, I_D vs V_{GS}, are shown in Fig. 4.4a, which also contains the linear response of V_{GS} as a function of pH. The sensitivity of the gate membrane can be calculated from the slope in the linear range, for a constant I_D value (see Fig. 4.4b). The Nernstian expected value for the sensitivity is 59.2 mV.pH^{-1} at 25 °C.

The most common commercial FETs used in SEGFET are the instrumentation amplifier AD620 (used as high impedance unity gain buffer), the operational amplifier LF356 (connected to the input pin of a readout circuit based on high input impedance J-FET operational amplifier, as unity gain buffer), the instrumentation amplifier with LT1167, and the commercial MOSFET CD4007UB [44, 80, 81].

The operational pH range of SEGFETs depends on the stability of the separative extended gate material [for example, it is known that SnO_2 fabricated via sol–gel is damaged in solutions with high pH (pH $>$ 9)] [39]. Table 4.2 compares the characteristics for some sensing gate films for FET-based sensors:

4.4 Recent Trends Using Field-Effect Sensors

Recent efforts have been focused on the design of field-effect sensors containing immobilized nanomaterials, which are suitable for electronic control and biological sensing [17, 18, 89]. The immobilization of nanomaterials including nanoparticles and nanotubes usually requires high-cost equipment and/or advanced manipulation techniques. One exception is the use of the LbL technique, through which manipulation with control at the molecular level can be achieved with experimental simplicity [4, 23–29]. The first studies reporting field-effect sensors containing LbL films were reported by Cui et al. [90, 91], in which poly(dimethyldiallylammonium

Table 4.2 Comparison of main parameters for some sensing gate films used in FET-based sensors

Sensing film	Preparation method	Device structure	Drift rate (mV/h)	Sensitivity (mV/pH)	pH range	Ref.
SiO_2	Thermally grown	ISFET	≤5	50–58	1–13	[82]
SnO_2	Sputtering	ISFET	9.1	58	2–12	[83]
SnO_2	Sol–gel	ISFET	6.73	57.4	1–9	[84]
ITO-PVP	Spin coating	EGFET	–	57–59	2–12	[85]
V_2O_5	Sol–gel	SEGFET	–	58.1	2–12	[38]
TiO_2:Ru	Sputtering	SEGFET	1.03	52.2	1–13	[72]
ITO	Commercial	SEGFET	–	58	2–12	[86, 87]
FTO	Commercial	SEGFET	–	50	2–12	[39]
Au-PVS/N-PANI	LbL	SEGFET	2.2	58	2–12	[88]

FTO fluorine-doped tin oxide films, *PVP* poly(4-vinylphenol), *Au-PVS/N-PANI* gold electrode modified with LbL film of poly(vinyl sulfonic acid) (PVS) and nanostructurated polyaniline (N-PANI)

chloride) (PDDA) was immobilized in the gate platform in conjunction with SnO_2 and SiO_2 nanoparticles. Since then, the use of LbL technique has been considered an efficient strategy to modify the gate platform in FET devices. This is the case of a biosensor for lactate detection developed by Jing-Juan Xu et al. with immobilization of MnO_2 nanoparticles alternated with lactate oxidase and PDDA on the gate of an ISFET. A better sensitivity and performance towards lactate detection was attributed to the nanostructured film modifying the gate [92].

Other 1D nanomaterials including nanowires and nanotubes have been reported as gate-modifying agents for enhanced sensitivity in FET devices. For example, Javey et al. reported an LbL assembly of nanowires (NW) building blocks for the fabrication of NW FETs using Ge/Si core–shell NWs as an approach for three-dimensional (3D) multifunctional electronics [93]. Poghossian et al. reported the first capacitive electrolyte-insulator semiconductor (EIS) structure using LbL films of poly(allylamine hydrochloride) (PAH) and PSS [58–60]. With the same strategy, Siqueira Jr. et al. proposed the first EIS capacitive sensor functionalized with an LbL film containing polyamidoamine (PAMAM) dendrimer and single-walled carbon nanotubes (SWNTs) with the enzyme penicillinase immobilized atop the film surface for detecting penicillin G [94, 95]. The high sensitivity achieved was attributed to the presence of SWNTs, thus representing a suitable platform for protein immobilization. The same film structure was used in a LAPS sensor [96]. For both modified EIS and LAPS devices, the film containing nanotubes enhanced the sensor performance with increased sensitivity towards penicillin G, also allowing for stable signals with low drift and fast response time.

Siqueira Jr. et al. also investigated the influence of these PAMAM/SWNT-Penicillinase film on the FET device performance, and associated the film morphology with the signal response [97]. It was demonstrated that LbL-based PAMAM/SWNTs films act as a membrane with two distinct functions: First, the

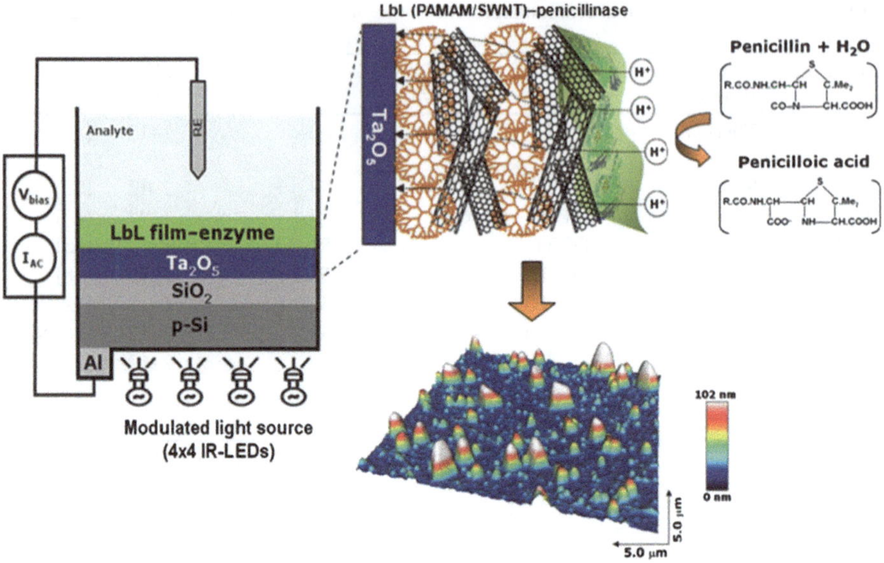

Fig. 4.5 Schematic representation of LAPS devices modified with PAMAM/SWNT LbL film containing a penicillinase layer absorbed on top; additionally the AFM image displays the morphology of the film-enzyme structure. Reprinted with permission from Ref. [98]. Copyright 2012 American Chemical Society

film allowed a stronger, more uniform adsorption of enzymes on the sensor surface. Second, the high porosity of the film, due to the interpenetration of nanotubes into dendrimer layers, facilitates the ion permeation resulting from enzymatic reactions through the film.

One important challenge to be faced in multiple sensing units in LAPS devices is to avoid cross-talk effects. Siqueira et al. solved this problem on LAPS biosensors modified with dendrimer-nanotubes using information visualization methods, in which projections techniques were implemented to treat the data [98]. Figure 4.5 shows a schematic representation of EIS and LAPS devices modified with PAMAM/SWNT LbL film containing a penicillinase layer absorbed atop.

The benefits of modifying EIS structures with LbL films to achieve biosensors with improved performance was also reported by Abouzar et al., who observed an amplification of the signal response upon alternating layers of polyelectrolytes and enzymes as gate membranes on the p-Si-SiO_2 EIS structure [99]. A new variant of EIS sensors has been produced, which comprised an array of individually addressable nanoplate field-effect capacitive biochemical sensors with an SOI (silicon-on-insulator) structure to determine pH and detect penicillin. It also allows for the label-free electrical monitoring of formation of polyelectrolyte multilayers and DNA (deoxyribonucleic acid)-hybridization event [100].

Another strategy to functionalize FEDs was demonstrated by Gun et al., modifying a capacitive EIS structure with gold nanoparticles and glucose oxidase, used as field-effect-based glucose biosensor. The co-immobilization of ferrocene redox species led to a two-fold increase in the biosensor sensitivity [101, 102].

Fernandes et al. described a gate membrane made with LbL films of dendrimers and phthalocyanine as SEGFET-based pH sensor, which was advantageous because metallophtalocyanines may act as artificial enzymes [44]. Semiconductor polymers have also been used as platforms for SEGFET sensing membranes. In a recent publication, nanostructure polyaniline LbL films were applied as modifiers gate membranes, exhibiting good physicochemical properties and near Nernstian sensitivity (58 mV.pH^{-1} with small voltage drift) [88]. The SEGFET sensor containing organic semiconductors also exhibited high stability within a pH range from 2 to 12 and linear pH sensitivity. Furthermore, a very low drift (an important feature for oxide sensitive membranes) and low response time (ca. 3 min) were observed. Other examples of field-effects sensors with different ways of modification can be found in refs. [5–7].

4.5 Final Remarks

FET-based devices have been proven as an efficient strategy for sensors and biosensors, mainly because of their facilitated fabrication with commercially available microelectronics components, which make it possible to produce devices in a large-scale at relatively low cost. Another advantage is the number of possible architectures leading to distinct devices including ISFETs, EGFETs, SEGFETs, EIS and LAPS, each of which exhibits advantages for specific applications.

Indeed, biosensing has benefited enormously from the development of FET sensor platforms, not only due to the design of specific FET architectures, but also because nanotechnological materials and techniques may be used to obtain gate platforms with tailored surfaces and functionalities. The latter features are crucial for improving the efficiency of biomolecules immobilization, leading to higher protein loadings, and as a consequence, better sensitivity and lower limit of detection.

Acknowledgments The authors are grateful to CNPq, FAPEMIG, FAPESP, and Rede nBioNet (CAPES) for the financial support.

References

1. Turner, A.P.F.: Biochemistry—biosensors sense and sensitivity. Science **290**, 1315–1317 (2000)
2. Kubik, T., Bogunia-Kubik, K., Sugisaka, M.: Nanotechnology on duty in medical applications. Curr. Pharm. Biotech. **6**, 17–33 (2005)

3. Vaseashta, A., Vaclavikova, M., Vaseashta, S., Gallios, G., Roy, P., Pummakarnchana, O.: Nanostructures in environmental pollution detection, monitoring, and remediation. Sci. Tech. Adv. Mater. **8**, 47–59 (2007)
4. Siqueira, Jr., J. R., Caseli, L., Crespilho, F. N., Zucolotto, V., Oliveira, Jr., O. N.: Biosens. Bioelectron. **25**, 1254 (2010)
5. Schoning, M.J.: "Playing around" with field-effect sensors on the basis of EIS structures. LAPS and ISFETs. Sensors **5**, 126–138 (2005)
6. Schoning, M.J., Poghossian, A.: Bio FEDs (field-effect devices): state-of-the-art and new directions. Electroanal **18**, 1893–1900 (2006)
7. Poghossian, A., Schöning, M.J.: Silicon-based chemical and biological field-effect devices. In: Grimes, C.A., Dichey, E.C., Pishko, M.V. (eds.) Encyclopedia of Sensors, vol. 9, pp. 463–533. American Scientific Publishers, Stevenson Ranch (2006)
8. Chaniotakis, N., Sofikiti, N.: Novel semiconductor materials for the development of chemical sensors and biosensors: a review. Anal. Chim. Acta **615**, 1–9 (2008)
9. Cui, Y., Wei, Q.Q., Park, H.K., Lieber, C.M.: Nanowire nanosensors for highly sensitive and selective detection of biological and chemical species. Science **293**, 1289–1292 (2001)
10. Keren, K., Berman, R.S., Buchstab, E., Sivan, U., Braun, E.: DNA-Templated Carbon Nanotube Field-Effect Transistor. Science **302**, 1380–1382 (2003)
11. Crespilho, F.N., Ghica, M.E., Florescu, M., Nart, F.C., Oliveira, Jr., O.N., Brett, C.M.A.: A strategy for enzyme immobilization on layer-by-layer dendrimer-gold nanoparticle electrocatalytic membrane incorporating redox mediator. Electrochem. Comm. **8**, 1665–1670 (2006)
12. Crespilho, F.N., Zucolotto, V., Brett, C.M.A., Oliveira, Jr., O.N., Nart, F.C.: Enhanced charge transport and incorporation of redox mediators in layer-by-layer films containing PAMAM-encapsulated gold nanoparticles. J. Phys. Chem. B **110**, 17478–17483 (2006)
13. Patolsky, F., Timko, B.P., Zheng, G.F., Lieber, C.M.: Nanowire-based nanoelectronic devices in the life sciences. Mrs Bulletin **32**, 142–149 (2007)
14. Merkoci, A.: Nanobiomaterials in electroanalysis. Electroanal **19**, 739–741 (2007)
15. Crespilho, F.N., Ghica, M.E., Gouveia-Caridade, C., Oliveira, Jr., O.N., Brett, C.M.A.: Enzyme immobilization on electroactive nanostructured membranes (ENM): Optimised architectures for biosensing. Talanta **76**, 922–928 (2008)
16. Crespilho, F.N., Lanfredi, A.J.C., Leite, E.R., Chiquito, A.J., 2009. Development of individual semiconductor nanowire for bioelectrochemical device at low overpotential conditions. Electrochem. Comm. **11**, 1744–1747 (2009)
17. Willner, I., Willner, B.: Biomolecule-Based Nanomaterials and Nanostructures. Nano Lett. **10**, 3805–3815 (2010)
18. Katz, E., Willner, I.: Biomolecule-functionalized carbon nanotubes: applications in nanobioelectronics. Chemphyschem **5**(8), 1085–1104 (2004)
19. Balasubramanian, K., Burghard, M.: Biosensors based on carbon nanotubes. Anal. Bioanal. Chem. **385**, 452–468 (2006)
20. Allen, B.L., Kichambare, P.D., Star, A.: Carbon nanotube field-effect-transistor-based biosensors. Adv. Mater. **19**, 1439–1451 (2007)
21. Kima, S.N., Rusling, J.F., Papadimitrakopoulos, F.: Carbon nanotubes for electronic and electrochemical detection of biomolecules. Adv. Mater. **19**, 3214–3228 (2007)
22. Siqueira, Jr., J.R., Gasparotto, L.H.S., Oliveira, Jr., O.N., Zucolotto, V.: Processing of electroactive nanostructured films incorporating carbon nanotubes and phthalocyanines for sensing. J. Phys. Chem. C **112**, 9050–9055 (2008)
23. Decher, G., Hong, J.D., Schmitt, J.: Buildup of ultrathin multilayer films by a self-assembly process 3. consecutively alternating adsorption of anionic and cationic polyelectrolytes on charged surfaces. Thin Solid Films **210**, 831–835 (1992)
24. Hammond, P.T.: Form and function in multilayer assembly: new applications at the nanoscale. Adv. Mater. **16**, 1271–1293 (2004)
25. Tangh, Z.Y., Wang, Y., Podsiadlo, P., Kotov, N.A.: Biomedical applications of layer-by-layer assembly: from biomimetics to tissue engineering. Adv. Mater. **18**, 3203–3224 (2006)

26. Ariga, K., Hill, J.P., Ji, Q.M.: Layer-by-layer assembly as a versatile bottom-up nanofabrication technique for exploratory research and realistic application. Phys. Chem. Chem. Phys. **9**, 2319–2340 (2007)
27. Lutkenhaus, J.L., Hammond, P.T.: Electrochemically enabled polyelectrolyte multilayer devices: from fuel cells to sensors. Soft Matter **3**, 804–816 (2007)
28. Quinn, J.F., Johnston, A.P.R., Such, G.K., Zelikin, A.N., Caruso, F.: Next generation, sequentially assembled ultrathin films: Beyond electrostatics. Chem. Soc. Rev. **36**, 707–718 (2007)
29. Ariga, K., Hill, J.P., Lee, M.V., Vinu, A., Charvet, R., Acharya, S.: Challenges and breakthroughs in recent research on self-assembly. Sci. Technol. Adv. Mater. **9**, 014109 (2008)
30. Ariga, K., Ji, Q.M., Hill, J.P.: Modern Techniques for Nano- and Microreactors-Reactions, p. 51, Springer-Verlag Berlin, Berlin (2010).
31. Li, M., Ishihara, S., Akada, M., Liao, M., Sang, L., Hill, J.P., Krishnan, V., Ma, Y., Ariga, K.: Electrochemical-coupling layer-by-layer (ECC-LbL) assembly. J. Am. Chem. Soc. **133**, 7348–7351 (2011)
32. Lu, W., Lieber, C.M.: Nanoelectronics from the bottom up. Nat. Mater. **6**, 841–850 (2007)
33. Bergveld, P.: Thirty years of ISFETOLOGY—What happened in the past 30 years and what may happen in the next 30 years. Sens. Actuat. B **88**, 1–20 (2003)
34. Poghossian, A.S.: Method of fabrication of ISFETs and CHEMFETs on a Si-SiO_2-Si structure. Sens. Actuat. B **13–14**, 653–654 (1993)
35. Matsuo, T., Esashi, M.: Method of ISFET fabrication. Sens. Actuat. B **1**, 77–96 (1981)
36. Liao, H.K., Chia, L.L., Chou, J.C., Chung, W.Y., Sun, T.P., Hsiung, S.K.: Study on pHpzc and surface potential of tin oxide gate ISFET. Mat. Chem. Phys. **59**, 6–11 (1999)
37. Batista, P.D., Mulato, M.: ZnO extended-gate field-effect transistors as pH sensors, Appl. Phys. Lett. **87**, 143508/1-3 (2005)
38. Guerra, E.M., Silva, G.R., Mulato, M.: Extended gate field effect transistor using V_2O_5 xerogel sensing membrane by sol-gel method. Solid State Sci. **11**, 456–460 (2009)
39. Batista, P.D., Mulato, M.: Polycrystalline fluorine-doped tin oxide as sensoring thin film in EGFET pH sensor. J. Mater. Sci. **45**, 5478–5481 (2010)
40. Sastre, A., Bassoul, P., Fretigny, C., Simon, J., Roger, J.P., Thami, T.: A mesomorphic amphiphilic phthalocyanine derivative used for the functionalization of the grid surface of a field effect transistor, New J. Chem. 569–578 (1988)
41. Hsu, H.Y., Wu, C.Y., Lee, H.C., Lin, J.L., Chin, Y.L., Sun, T.P.: Sodium and potassium sensors based on separated extended gate field effect transistor. Biomed. Eng. Appl. Basis Commun. **21**, 441–444 (2009)
42. Lin, J.L., Hsu, H.Y.: Study of sodium ion selective electrodes and differential structures with anodized indium tin oxide. Sensors **10**, 1798–1809 (2010)
43. Xue, W., Cui, T.: A thin-film transistor based acetylcholine sensor using self-assembled carbon nanotubes and SiO_2 nanoparticles. Sens. Act. B **134**, 981–987 (2008)
44. Fernandes, E.G.R., Vieira, N.C.S., de Queiroz, A.A.A., Guimarães, F.E.G.: Immobilization of poly(propylene imine) dendrimer/Nickel phthalocyanine as nanostructured multilayer films to be used as gate membranes for SEGFET pH sensors. J. Phys. Chem. C **114**, 6478–6483 (2010)
45. Vieira, N.C.S., Figueiredo, A., de Queiroz, A.A.A., Zucolotto, V., Guimarães, F.E.G.: Self-assembled of dendrimers and metallophthalocyanines as FET-based glucose biosensors. Sensors **11**, 9442–9449 (2011)
46. Jaffrezic-Renault, N.: New trends in biosensors for organophosphorus pesticides. Sensors **1**, 60–74 (2001)
47. Lee, H.C., Wu, W.Y., Lin, J.L., Chin, Y.L., Lee, K.Y., Sun, T.P.: Evolution of the TiO_2 membrane on ITO PET substrate applied to a lactate biosensor using potentiometric differential readout circuit. IEEE Sens. Conf. 898–901 (2008)

48. Lee, S.W., Kim, B.S., Chen, S., Shao-Horn, Y., Hammond, P.T.: Layer-by-layer assembly of all carbon nanotube ultrathin films for electrochemical applications. J. Am. Chem. Soc. **131**, 671–679 (2009)
49. Siqueira, Jr., J.R., Crespilho, F.N., Zucolotto, V., Oliveira, Jr., O.N.: Bifunctional electroactive nanostructured membranes. Electrochem. Comm. **9**, 2676–2680 (2007)
50. Lvov, Y., Ariga, K., Ichinose, I., Kunitake, T.: Assembly of multicomponent protein films by means of electrostatic layer-by-layer adsorption. J. Am. Chem. Soc. **117**, 6117–6123 (1995)
51. Zucolotto, V., Daghastanli, K.R.P., Hayaaka, C.O., Riul, Jr., A., Ciancaglini, P., Oliveira, Jr., O. N.: Using capacitance measurements as the detection method in antigen-containing layer-by-layer films for biosensing. Anal. Chem. **79**, 2163–2167 (2007)
52. Mertens, J., Rogero, C., Calleja, M., Ramos, D., Martin-Gago, J.A., Briones, C., Tamayo, J.: Label-free detection of DNA hybridization based on hydration-induced tension in nucleic acid films. Nat. Nanotech. **3**, 301–307 (2008)
53. Siqueira, Jr., J.R., Gasparotto, L.H.S., Crespilho, F.N., Carvalho, A.J.F., Zucolotto, V., Oliveira, Jr., O.N.: Physicochemical properties and sensing ability of metallophthalocyanines/chitosan nanocomposites. J. Phys. Chem. B **110**, 22690–22694 (2006)
54. Krämer, M., Pita, M., Zhou, J., Ornatska, M., Poghossian, A., Schöning, M.J., Katz, E.: Coupling of biocomputing systems with electronic chips: electronic interface for transduction of biochemical information. J. Phys. Chem. C **113**, 2573–2579 (2009)
55. Zucolotto, V., Pinto, A.P.A., Tumolo, T., Moraes, M.L., Baptista, M.S., Riul, Jr., A., Araujo, A.P.U., Oliveira, Jr., O.N.: Catechol biosensing using a nanostructured layer-by-layer film containing Cl-catechol 1,2-dioxygenase. Biosens. Bioelectron. **21**, 1320–1326 (2006)
56. Perinotto, A.C., Caseli, L., Hayasaka, C.O., Riul, Jr., A., Oliveira, Jr., O.N., Zucolotto, V.: Dendrimer-assisted immobilization of alcohol dehydrogenase in nanostructured films for biosensing: Ethanol detection using electrical capacitance measurements. Thin Solid Films **516**, 9002–9005 (2008)
57. Crespilho, F.N., Iost, R.M., Travain, S.A., Oliveira, Jr., O.N., Zucolotto, V.: Enzyme immobilization on Ag nanoparticles/polyaniline nanocomposites. Biosens. Bioelectron. **24**, 3073–3077 (2009)
58. Poghossian, A., Abouzar, M.H., Amberger, F., Mayer, D., Han, Y., Ingebrandt, S., Offenhausser, A., Schoning, M.J.: Field-effect sensors with charged macromolecules: Characterisation by capacitance-voltage, constant-capacitance, impedance spectroscopy and atomic-force microscopy methods. Biosens. Bioelectron. **22**, 2100–2107 (2007)
59. Poghossian, A., Abouzar, M.H., Sakkari, M., Kassab, T., Han, Y., Ingebrandt, S., Offenhausser, A., Schoning, M.J.: Field-effect sensors for monitoring the layer-by-layer adsorption of charged macromolecules. Sens. Actuat. B **118**, 163–170 (2006)
60. Poghossian, A., Ingebrandt, S., Abouzar, M.H., Schoning, M.J.: Label-free detection of charged macromolecules by using a field-effect-based sensor platform: Experiments and possible mechanisms of signal generation. Appl. Phys. A-Mater. **87**, 517–524 (2007)
61. Wagner, T., Rao, C., Kloock, J.P., Yoshinobu, T., Otto, R., Keusgen, M., Schöning, M.J.: "LAPS Card"—A novel chip card-based light-addressable potentiometric sensor (LAPS). Sens. Actuat. B **118**, 33–40 (2006)
62. Wagner, T., Yoshinobu, T., Rao, C.W., Otto, R., Schöning, M.J.: "All-in-one" solid-state device based on a light-addressable potentiometric sensor platform. Sens. Actuat. B **117**, 472–479 (2006)
63. Van Der Spiegel, J., Lauks, I., Chan, P., Babic, D.: The extended gate chemical sensitive field effect transistor as multi-species microprobe. Sens. Actuat. B **4**, 291–298 (1983)
64. Yate, D.E., Levine, S., Healy, T.W.: Site-binding model of the electrical double layer at the oxide/water interface. J. Chem. Soc. Faraday Trans. **1**(70), 1807–1818 (1974)
65. A. J. Bard and Faulkner, Electrochemical methods fundamentals and applications, John Wiley & Sons, New York, 1980.

66. Chou, J.C., Kwan, P.K., Chen, Z.J.: SnO_2 Separative Structure Extended Gate H^+-Ion Sensitive Field Effect Transistor by the Sol–Gel Technology and the Readout Circuit Developed by Source Follower. Jpn. J. Appl. Phys. **42**, 6790–6794 (2003)
67. Janata, J.: Electrochemistry of chemically sensitive field effect transistors. Sens. Actuat. B **4**, 255–265 (1983)
68. Yin, L.T., Chou, J.C., Chung, W.Y., Sun, T.P., Hsiung, S.K.: Study on all-solid-state chloride sensor based on tin oxide/indium tin oxide glass. Jpn. J. Appl. Phys. **50**, 037001–037009 (2011)
69. Batista, P.D., Mulato, M., Graeff, C.F.O., Fernandez, F.J.R., Marques, F.D.: SnO_2 extended gate field-effect transistor as pH sensor. Braz. J. Phys. **36**, 478–481 (2006)
70. Guerra, E.M., Mulato, M.: Synthesis and characterization of vanadium oxide/hexadecylamine membrane and its application as pH-EGFET sensor. J. Sol–Gel Sci. Technol. **52**, 315–320 (2009)
71. Guidelli, E.J., Guerra, E.M., Mulato, M.: Ion sensing properties of vanadium/tungsten mixed oxides. Mat. Chem. Phys. **125**, 833–837 (2011)
72. Chou, J.C., Chen, C.W.: Long-term monitor of seawater by using TiO_2:ru sensing electrode for hard clam cultivation. World Acad. Sci. Eng. Technol. **53**, 349–353 (2009)
73. Jan, S.S., Chiang, J.L., Chen, Y.C., Chou, J.C., Cheng, C.C.: Characteristics of the hydrogen ion-sensitive field effect transistors with sol–gel-derived lead titanate gate. Anal. Chim. Acta. **469**, 205–216 (2002)
74. Chen, J.C., Chou, J.C., Sun, T.P., Hsiung, S.K.: Portable urea biosensor based on the extended-gate field effect transistor. Sens. Actuat. B **91**, 180–186 (2003)
75. Yin, L.T., Lin, Y.T., Leu, Y.C., Hu, C.Y.: Enzyme immobilization on nitrocellulose film for pH-EGFET type biosensors. Sens. Actuat. B **148**, 207–213 (2010)
76. Yin, L.T., Chou, J.C., Chung, W.Y., Sun, T.P., Hsiung, K.P., Hsiung, S.K.: Glucose ENFET doped with MnO_2 powder. Sens. Actuat. B **76**, 187–192 (2001)
77. Ishige, Y., Shimoda, M., Kamahori, M.: Extended-gate FET-based enzyme sensor with ferrocenyl-alkanethiol modified gold sensing electrode. Biosens. Bioelectron. **24**, 1096–1102 (2009)
78. Ishige, Y., Shimoda, M., Kamahori, M.: Immobilization of DNA Probes onto gold surface and its application to fully electric detection of DNA hybridization using field effect transistor sensor. Jpn. J. Appl. Phys. **45**, 3776–3783 (2006)
79. Kim, D.S., Jeong, Y.T., Park, H.J., Shin, J.K., Choi, P., Lee, J.H., Lim, G.: An FET-type charge sensor for highly sensitive detection of DNA sequence. Biosens. Bioelectron. **20**, 69–74 (2004)
80. Chi, L.L., Yin, L.T., Chou, J.C., Chung, W.Y., Sun, T.P., Hsiung, K.P., Hsiung, S.K.: Study on separative structure of EnFET to detect acethylcoline. Sens. Actuat. B **71**, 68–72 (2000)
81. Chi, L.L., Chou, J.C., Chung, W.Y., Sun, T.P., Hsiung, S.K.: Study on extended gate field effect transistor with tin oxide sensing membrane. Mat. Chem. Phys. **63**, 19–23 (2000)
82. Castellarnau, M., Zine, N., Bausells, J., Madrid, C., Juárez, A., Samitier, J., Errachid, A.: ISFET-based biosensor to monitor sugar metabolism in bacteria. Mat. Sci. Eng. C **28**, 680–685 (2008)
83. Chou, J.C., Wang, Y.F.: Temperature characteristics of a-Si:H gate ISFET. Mater. Chem. Phys. **70**, 107–111 (2001)
84. Chou, J.C., Wang, Y.F.: Preparation and study on the drift and hysteresis properties of the tin oxide gate ISFET by the sol-gel method. Sens. Actuat. B **86**, 58–62 (2002)
85. Nguyen, T.N.T., Seol, Y.G., Lee, N.E.: Organic field-effect transistor with extended indium tin oxide gate structure for selective pH sensing. Organic Electronics **12**, 1815–1821 (2011)
86. Yin, L.T., Chou, J.C., Chung, W.Y., Sun, T.P., Hsiung, S.K.: Separate structure extendet gate H^+-ion sensitive field effect transistor on a glass substrate. Sens. Actuat. B **71**, 106–111 (2000)
87. Yin, L.T., Chou, J.C., Chung, W.Y., Sun, T.P., Hsiung, S.K.: Study of indium tin oxide thin film for separative extended gate ISFET. Mat. Chem. Phys. **70**, 12–16 (2001)

88. Vieira, N.C.S., Fernandes, E.G.R., Faceto, A.D., Zucolotto, V., Guimarães, F.E.G.: Nanostructurated polyaniline thin films as pH sensing membranes in FET-based devices. Sens. Actuat. B **160**, 312–317 (2011)
89. Daniel, M.C., Astruc, D.: Gold nanoparticles: assembly, supramolecular chemistry, quantum-size-related properties, and applications toward biology, catalysis, and nanotechnology. Chem. Rev. **104**, 293–346 (2004)
90. Cui, T.H., Hua, F., Lvov, Y.: FET fabricated by layer-by-layer nanoassembly. IEEE Trans. Elec. Dev. **51**, 503–506 (2004)
91. Cui, T.H., Liu, Y., Zhu, M.: Field-effect transistors with layer-by-layer self-assembled nanoparticle thin films as channel and gate dielectric. Appl. Phys. Lett. **87**, 183105–183105-3 (2005)
92. Xu, J.J., Zhao, W., Luo, X.L., Chen, H.Y.: A sensitive biosensor for lactate based on layer-by-layer assembling MnO_2 nanoparticles and lactate oxidase on ion-sensitive field-effect transistors. Chem. Comm. 792–794 (2005)
93. Javey, A., Nam, S., Friedman, R.S., Yan, H., Lieber, C.M.: Layer-by-layer assembly of nanowires for three-dimensional, multifunctional electronics. Nano Lett. **7**, 773–777 (2007)
94. Siqueira, Jr., J.R., Abouzar, M.H., Backer, M., Zucolotto, V., Poghossian, A., Oliveira, Jr., O.N., Schoning, M.J.: Carbon nanotubes in nanostructured films: Potential application as amperometric and potentiometric field-effect (bio-)chemical sensors. Phys. Stat. Sol. A **206**, 462–467 (2009)
95. Siqueira, Jr., J.R., Abouzar, M.H., Poghossian, A., Zucolotto, V., Oliveira, Jr., O.N., Schöning, M.J.: Penicillin biosensor based on a capacitive field-effect structure functionalized with a dendrimer/carbon nanotube multilayer. Biosens. Bioelectron. **25**, 497–501 (2009)
96. Siqueira, Jr., J.R., Werner, C.F., Backer, M., Poghossian, A., Zucolotto, V., Oliveira, Jr., O.N., Schoning, M.J.: Layer-by-Layer Assembly of Carbon Nanotubes Incorporated in Light-Addressable Potentiometric Sensors. J. Phys. Chem. C **113**, 14765–14770 (2009)
97. Siqueira, Jr., J. R.; Bäcker, M.; Poghossian, A.; Zucolotto, V.; Oliveira, Jr., O. N.; Schöning, M. J.: Associating biosensing properties with the morphological structure of multilayers containing carbon nanotubes on field-effect devices. Physica Status Solidi A-Applications and Materials Science **207**, 781–786 (2010)
98. Siqueira, Jr., J. R.; Maki, R. M.; Paulovich, F. V.; Werner, C. F.; Poghossian, A.; De Oliveira, M. C. F.; Zucolotto, V.; Oliveira, Jr., O. N.; Schöning, M. J.: Use of information visualization methods eliminating cross talk in multiple sensing units investigated for a light-addressable potentiometric sensor. Anal. Chem. **82**, 61–65 (2010)
99. Abouzar, M. H., Siqueira, Jr., J. R., Poghossian, A., Oliveira, Jr., O. N., Moritz, W., Schöning, M.J.: Capacitive electrolyte-insulator-semiconductor structures functionalised with a polyelectrolyte-enzyme multilayer: new strategy for enhanced field-effect biosensing. Phys. Stat. Sol. A **207**, 884–890 (2010)
100. Abouzar, M.H., Poghossian, A., Pedraza, A.M., Gandhi, D., Ingebrandt, S., Moritz, W., Schöning, W.J.: An array of field-effect nanoplate SOI capacitors for (bio-)chemical sensing. Biosens. Bioelectron. **26**, 3023–3028 (2011)
101. Gun, J., Schöning, M.J., Abouzar, M.H., Poghossian, A., Katz, E.: Field-effect nanoparticle-based glucose sensor on a chip: amplification effect of coimmobilized redox species. Electroanal **20**, 1748–1753 (2008)
102. Gun, J., Gutkin, V., Lev, O., Boyen, H.G., Saitner, M., Wagner, P., D'Olieslaeger, M., Abouzar, M.H., Poghossian, A., Schöning, M.J.: Tracing gold nanoparticle charge by electrolyte-insulator-semiconductor devices. J. Phys. Chem. C **115**, 4439–4445 (2011)

Chapter 5
Using Supramolecular Chemistry Strategy for Mapping Electrochemical Phenomena on the Nanoscale

Anna Thaise Bandeira Silva, Janildo Lopes Magalhães, Eduardo Henrique Silva Sousa and Welter Cantanhêde da Silva

Abstract The main goal of this chapter is to show how the supramolecular chemistry strategy is used to map electrochemical phenomena at the nanoscale of low- dimensional highly organized hybrid structures containing several building blocks such as metallic nanoparticles, carbon nanotubes, metallic phthalocyanine, (bio)polymers, enzymes and synthetic polymers. In this sense, the principles of supramolecular chemistry as constitutional dynamic character of the reactions, functional recognition, and self-organization are explored from interaction between biomolecules and several supramolecular architectures in order to modulate the physicochemical properties that arise at molecular level. The developed platforms with high control of these electrochemical properties become interesting devices for sensor and biosensor applications. Additionally, we describe nature-made biological nanosensors as an inspirational scaffold that might lead us to create advanced novel material as well.

Abbreviations

AA	Ascorbic acid
ADA	Adamantine
ATP	Adenosine triphosphate
AuNPs	Gold nanoparticles
CD	Cyclodextrin
CDC	Constitutional dynamic chemistry
Chit	Chitosan

A. T. B. Silva · J. L. Magalhães · W. C. da Silva (✉)
Departamento de Química, Centro de Ciências da Natureza, Universidade Federal do Piauí, Teresina, PI 64049-550, Brazil
e-mail: welter@ufpi.edu.br

E. H. S. Sousa
Departamento de Química Orgânica e Inorgânica, Centro de Ciências, Universidade Federal do Ceará, Fortaleza, CE 60455-760, Brazil

F. N. Crespilho (ed.), *Nanobioelectrochemistry*, DOI: 10.1007/978-3-642-29250-7_5,

CMC	Carboxymethylcellulose
CNT	Carbon nanotubes
CO	Carbon monoxide
CooA	CO heme-based sensor
CoTsPc	Cobalt (II) tetrasulfonated phthalocyanine
Crown-C_{60}	Benzo-18-crown-6 fullerene
CVs	Cyclic voltammograms
CysSH	Cysteine
$C_{18}H_{37}SH$	N-octadecylmercaptan
C60	Fullerene
DA	Dopamine
DAQ	Dopamine quinine
DevS	Oxygen heme-based sensor from M. tuberculosis
DMPA	Dimyristoyl phosphatidic acid
DNA	Deoxyribonucleic acid
DosT	Oxygen heme-based sensor from M. tuberculosis
E-AB	Electrochemical aptamer-based
ENM	Electroactive nanostructured membranes
FePc	Iron phthalocyanine
FixL	Oxygen sensor histidine kinase protein found mainly in Rhizobia
FixJ	Response regulator protein found mainly in Rhizobia
Fe_3O_4-NPs	Fe_3O_4-nanoparticles
GAF	Regulatory domain named after the proteins cGMP-specific phosphodiesterases, adenylyl cyclase and FhlA
GNF	Graphene nanosheet films
GOX	Glucose oxidase
HNOB	Heme NO-binding domain
HV^+	Viologen
HV^{2+}	Hexyl viologen dication
H_2O_2	Hydrogen peroxide
ITO	Indium tin oxide
LB	Langmuir–Blodgett
LBL	Layer-by-layer
LEDs	Light-emitting diodes
Mtb	Mycobacterium tuberculosis
MWCNTs	Multi-walled carbon nanotubes
NADH	Nicotinamide adenine dinucleotide
NO	Nitric oxide
NPAS2	Neuronal PAS domain 2, mammalian transcription factor
O_2	Oxygen
PAH	Poly(allylamine hydrochloride)
PAMAM	Polyamidoamine dendrimer
PAS	Sensor domain named after the eukaryotic proteins period, Arnt and Single-minded

PB	Prussian blue
PB-CD NPs	Prussian blue nanoparticles protected by β-cyclodextrin
PB-NPs	Prussian blue nanoparticles
PBS	Phosphate buffer solution
PVP	Polyvinylsulfonate
Pyr-NH_3^+	Alkylammonium pyrene
rGO	Reduced graphene oxide
rMe	Redox mediator
RNA	Ribonucleic acid
SAM	Self-assembled monolayer
SCE	Saturated calomel electrode
SCHIC	Sensor containing heme instead of cobalamin domain
SEM	Scanning electronic microscopy
sGC	Soluble guanylate cyclase
SPE	Screen-printed electrode
SWCNTs	Single-walled carbon nanotubes
UA	Uric acid

5.1 General Overview

This chapter focuses on the concepts, strategies for self-assembly and current development of supramolecular chemistry regarding sensors and biosensor construction [1]. Supramolecular chemistry is primarily involved in the understanding and interpretation of new molecular phenomena that takes place at the nanoscale. This is an excellent bottom-up approach to nanoscience and nanotechnology [2, 3]. It is mainly based on the molecular recognition and self-organization of components that interact by several spontaneous secondary interactions such as electrostatic force, hydrogen bonding, dipole-dipole, charge transfer, π-π stacking interactions and metal ion coordination [2, 4]. The choice of building blocks plays a key role in the construction of new functional supramolecular entities. These species also undergo continuous modification in their constitutions until adduct formation. Since both reaction and conformational dynamics are involved in the self-assembly process, there is a continuous modification of the supramolecular environment allowing molecular rearrangement of its various components leading to the same final entity or other ones [2, 5–8]. This supramolecular self-assembly is known as constitutional dynamic chemistry (CDC) and has been explored to design complex systems with or without a specific biomolecule [2]. The construction of new nanostructures and functional nanoplatforms reported here is only carried out through bottom up approaches including co-precipitation [7, 9], self-assembled monolayer (SAM) [10], Langmuir–Blodgett (LB) [11] and layer-by-layer (LbL) [12–15] techniques. These entities have been assembled employing several building block such as metallic nanoparticles [16], carbon nanotubes [8], metallic phthalocyanine [5–7], (bio)polymers [8, 17], semiconductor quantum dots [9, 18], graphene [10], enzymes [19, 20], DNA [1, 21] and antigen–antibody pairs [9, 22]. Here we have emphasized how the control of several physical and chemical properties can be achieved by using supramolecular approaches to develop advanced functional nanomaterials as well as nanodevices for specific purposes [23]. Additionally, we have briefly presented some examples of nature-made biological sensors as a way to inspire bionanoengineering of highly sophisticated biological sensors in the near future.

5.2 Construction of Nanoplatforms and Supramolecular Electrochemistry from Functional Electrodes

The design, production and control of new nanomaterials and functional nanodevices take into account the properties (e.g., functional groups and geometry) of each component that interacts at molecular level [24–26]. For example, β-cyclodextrin (CD) is a typical cyclic oligosaccharide with a chemical structure containing a hydrophobic inner cavity and a hydrophilic outer surface. This kind of species has been used to prepare inclusion complexes as well as supramolecular architectures [27]. Also, prussian blue complex (PB) is a well-known mixed valence compound with iron (II) and iron (III) coordinated to cyanide ligands used as DNA probe, catalyst and magnet [28, 29]. Supramolecular self-assembly of mono- or multilayers of prussian blue nanoparticles (PB-NPs) protected by β-CD and poly(allylamine hydrochloride) (PAH) onto indium tin oxide (ITO) is an attractive and efficient method for drug delivery. Recently, Silva and co-workers [30] reported the formation of a compact microcubic structures by electrostatic interactions of oppositely charged PB-NPs and CD molecules as suggested by scanning electronic microscopy (SEM) (Fig. 5.1a). Deposition of PB nanoparticles protected by β-CD (PB–CD NPs) onto anionic polyelectrolyte alternately with PAH onto ITO substrate produced the nanostructured LbL platform (PAH/PB–CD) that shows high electrochemical response (Fig. 5.1b) and stability with regards to host–guest studies.

Novel supramolecular strategies have been reported based on the precise organization of nanomaterials [23]. Carbon nanotubes (CNT), including single-walled (SWCNTs) or multi-walled ones (MWCNTs), represent a class of nanomaterials with unique characteristics such as amazing mechanical and electrical properties, high surface area and high performance for sensor and biosensor applications [31–33]. The combination of CNT with other (bio)materials such as enzyme, inorganic compounds and fullerene (C60) produced new functional platforms. For example, due to the low biological affinity exhibited by some microelectrodes, CD electropolymerization onto screen-printed electrode (SPE) supported with multiwalled carbon nanotubes has been carried out as an enzyme

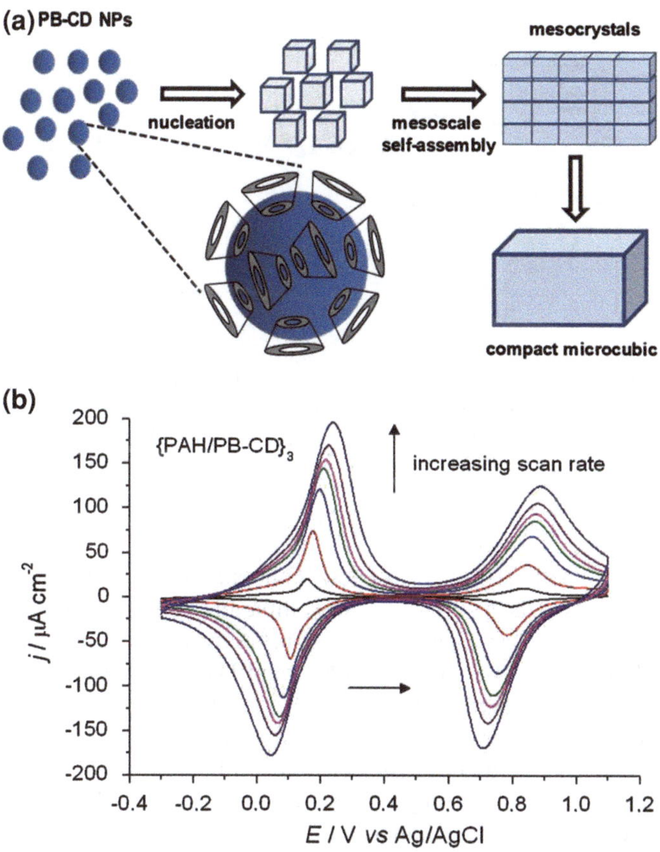

Fig. 5.1 **a** For compact microcubic structure formation, the PB–CD nanoparticles undergo nucleation process within the LbL flask conducting to the formation of microcrystals that support a mesoscale self-assembly process and a final supramolecular conversion to compact microcubic structures. **b** Cyclic voltammograms (CVs) for self-assembly {PAH/PB–CD} multilayers onto ITO electrode containing three bilayers at various scan rate: 10–200 mV s^{-1}. Electrolyte: KCl—0.2 mol L^{-1}, T = 25 °C. Adapted with permission from [30]

entrapment strategy [19] (Fig. 5.2). In this case the supramolecular interaction between CD polymer and MWCNTs is responsible for creating a biocompatible environment for glucose oxidase enzyme (GOX) immobilization, allowing a quick and easy detection of dopamine (DA) as suggested by electrochemical analysis.

In another report using the CDC approach, functional modulation was achieved by controlling LbL film but without a specific biomolecule. In order to achieve this goal, Luz and co-workers [8] assembled two LbL platforms that take advantage of alternate immobilization of cobalt (II) tetrasulfonated phthalocyanine ($Co^{II}TsPc$), chitosan (Chit) and SWCNTs. $Co^{II}TsPc$ belongs to a group of metallic phthalocyanines that show delocalized π-electrons and planar geometry with unique

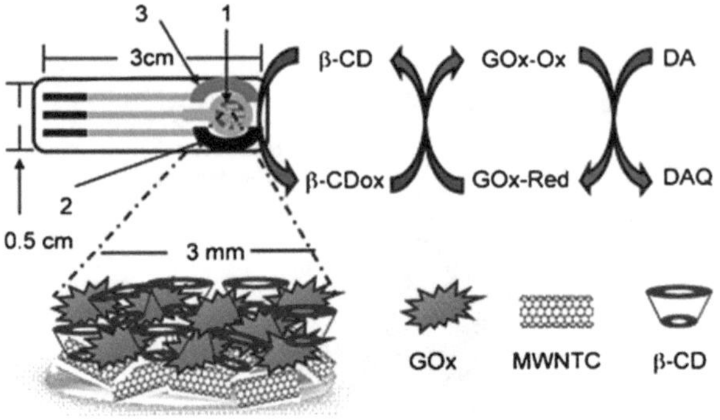

Fig. 5.2 Schematic representation of the screen-printed electrode modified with MWCNTs containing CD and GOX. The SPE contains: **a** working electrode, **b** counter electrode and **c** reference electrode. The dopamine detection is proposed from conversion of dopamine (DA) to dopamine quinine (DAQ) by GOX using β-CD as redox mediator. Reprinted with permission from [19]

properties as well defined redox couples, high catalytic activity and thermal stability [5–7]. Chitosan is a nontoxic, biocompatible and biodegradable polysaccharide, widely used for metal adsorption, delivery system and biomedical applications [34, 35]. ITO-{Chit/Co^{II}TsPc}$_n$ and ITO-{Chit-SWCNTs/Co^{II}TsPc}$_n$ architectures demonstrated that the intimate contact of the adjacent Co^{II}TsPc and SWCNTs layers are responsible for a supramolecular charge transfer. This effect promotes an increase of faradaic currents and film stability. As a result, biomolecules could be detected through their interactions with ITO-{Chit-SWCNTs/Co^{II}TsPc} electrode using CDC analysis.

Electron transfer reactions are key processes responsible for the maintenance of life. Certainly, supramolecular principles can help our understanding of the mechanisms of many biological processes such as photosynthetic reactions, oxidative phosphorylation, and many other events such those observed in the respiratory chain [1, 4]. Non-covalent functionalization of CNT has attracted investigation in technological applications as photovoltaic cells and light-emitting diodes (LEDs).

One advantage is to preserve the electronic structure of CNT, but it is also severely limited by chemical and thermal damage to the tubes [4]. Supramolecular nanohybrids are obtained from both π-π stacking of pyrene on the SWCNTs surface, and alkyl ammonium-crown ether interactions. The procedure used for self-assembling these sophisticated SWNTs-C_{60} nanohybrids using functionalized alkylammonium pyrene (Pyr-NH_3^+) and benzo-18-crown-6 fullerene (crown-C_{60}) both involve the solubilization of carbon nanotubes and maintenance of the electronic structure of Pyr-NH_3^+. Figure 5.3 illustrates a photoinduced electron transfer process in a self-assembled SWCNTs-C_{60} hybrid with SWCNTs and crown-C_{60} acting as an electron donor and acceptor, respectively [4]. This SWNT/Pyr-NH_3^+/crown-C_{60} system plays a role in promoting electron transfer and

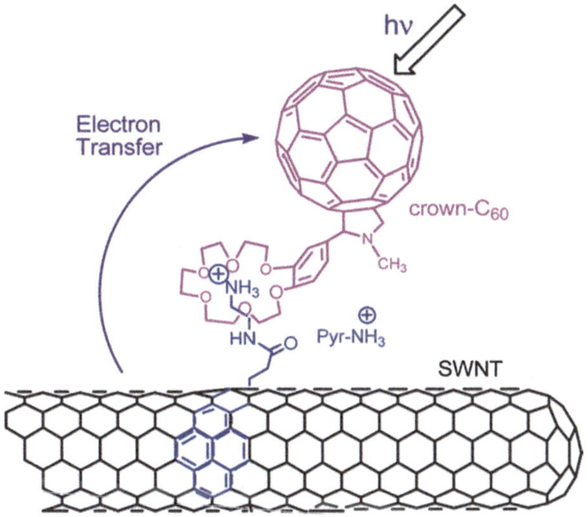

Fig. 5.3 Schematic illustration for SWNTs/Pyr-NH_3^+/C_{60} nanohybrids system with SWCNTs and fullerene acting as electron donor and acceptor, respectively. Reprinted with permission from [4]

converting hexyl viologen dication (HV^{2+}) to reduced viologen (HV^+). It is believed that self-assembled nanohybrid systems might have applications as artificial photosynthetic devices.

Graphene is a two-dimensional material with a hexagonal arrangement of carbon atoms, which has caught increasing scientific interest due to its high surface area, excellent electronic conductivity, high mechanical strength, thermal stability and ease of functionalization [36, 37]. Graphene-based materials exhibit appealing physical and chemical properties for use in enzymatic electron transfer processes and catalytic conversion of small biomolecules such as hydrogen peroxide (H_2O_2), ethanol and NADH. By controlled adsorption of reduced graphene oxide (rGO) onto SAM of n-octadecylmercaptan ($C_{18}H_{37}SH$) gold electrodes, Yang et al. [10] developed an effective method to produce graphene nanosheet films (GNF). The duration of rGO dispersion immobilization by SAM onto electrode was controlled in order to obtain a well defined thickness for better charge transport as supported by impedance spectroscopy and cyclic voltammetry analysis (Fig. 5.4). Besides this, the GNF/SAM modified electrode showed excellent electrocatalytical performance toward ascorbic acid (AA), dopamine and uric acid (UA) including simultaneous determination of these analytes (Fig. 5.4). It is reasonable to expect that modified electrodes containing graphene will become a very attractive nanomaterial for the development of new sensor and biosensors devices [38].

Supramolecular electrochemistry of functional nanoplatforms is a new multidisciplinary approach for nanoscience and nanotechnology devoted to investigate electrochemical phenomena by the combinations (or electrodes manipulation) and observation of direct electrochemical analysis of nanomaterial. This combination allows studying complex entities of interest for sensor and biosensor applications.

Cysteine (CysSH) is an important sulfur-containing amino acid that plays fundamental roles in biological systems [39] and it is also widely employed as a

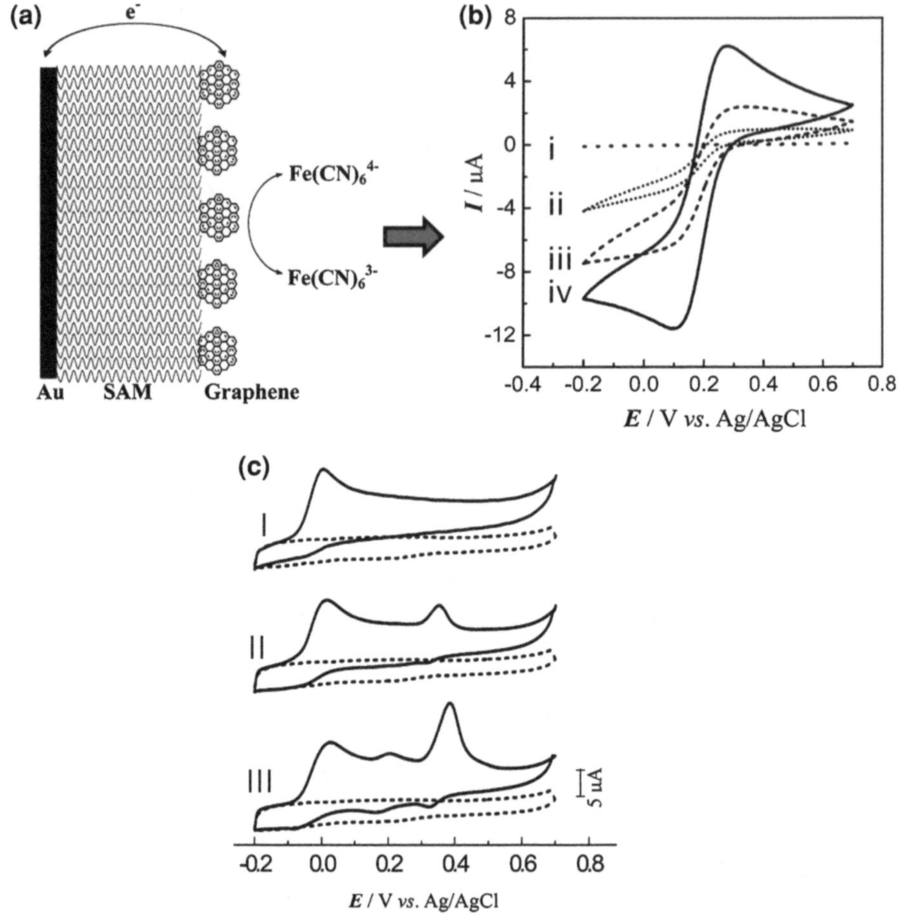

Fig. 5.4 **a** Schematic illustration for graphene-based nanomaterial; **b** CVs for GNF/SAM modified electrode in 0.50 mol L^{-1} KCl of 5.0 m mol L^{-1} hexacyanoferrate(III) after immersing the SAM electrode into a rGO dispersion in N,N-Dimethylformamide (1 mg/mL) during several times—0 (curve i) to 120 min (*curve iv*), scan rate = 50 mV s^{-1}; **c** CVs for GNF/SAM platform in 0.10 mol L^{-1} phosphate buffer solution (PBS) (pH 7.0): I without (*dotted line*) or with (*solid line*) 0.5 m mol L^{-1} AA, II 0.5 m mol L^{-1} AA and 0.25 m mol L^{-1} UA and III 0.5 m mol L^{-1} AA, 0.25 m mol L^{-1} UA, and 0.05 m mol L^{-1} DA. Adapted with permission from [10]

food supplement, pharmaceutical drug and treatment of skin damage. Since the thiol group of CysSH is extremely reactive and involved in a great number of biochemical reactions, there is significant interest to understand direct oxidation of thiol groups onto nanostructured electrodes. Conventional modified electrodes including glassy carbon and gold substrates show a slow thiol oxidation reaction and overpotential [40, 41]. Based on this, Santos and co-workers [5] investigated the influence of supramolecular organization of a Co^{II}TsPc complex assembled

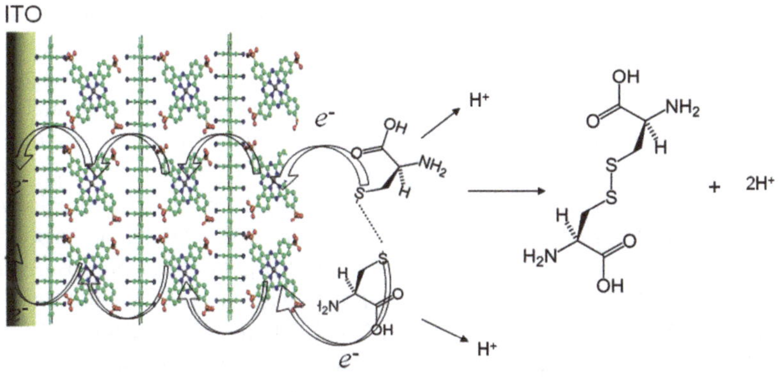

Fig. 5.5 Schematic representation of the cysteine oxidation mediated by modified LbL electrode containing 3-bilayers of the insulator PAH and $Co^{III}TsPc$ species. These supramolecular layers are electrically connected working through an electron hopping mechanism promoted by the redox sites of the $Co^{III}TsPc$ species right after oxidation of cysteine to cystine. Reproduced with permission from [5]

with insulator PAH in hybrid nanostructured electrodes. As a result, the supramolecular environment of $Co^{II}TsPc$ species showed a remarkable influence on the redox properties exhibited by ITO-{PAH/$Co^{II}TsPc$}$_3$ electrode and cysteine catalytic oxidation. Nanostructured $Co^{II}TsPc$ electrode showed high electrochemical stability and two redox processes with $E_{1/2}$ values at −0.72 e −0.58 V *versus* saturated calomel electrode (SCE) assigned to the redox pairs $[TsPc]^{6-}$/ $[TsPc]^{5-}$ and [$Co^{I}TsPc^{5-}$/$Co^{II}TsPc^{4-}$], respectively, and an irreversible process centered at 0.40 V corresponding to [$Co^{II}TsPc^{4-}$/$Co^{III}TsPc^{3-}$], according to the following equations:

$$[\mathrm{TsPc}]^{6-} \rightarrow [\mathrm{TsPc}]^{5-} + \mathrm{e}^-(E_{\mathrm{pa1}}) \quad (1)$$

$$[\mathrm{Co^{I}TsPc}]^{5-} \rightarrow [\mathrm{Co^{II}TsPc}]^{4-} + \mathrm{e}^-(E_{\mathrm{pa2}}) \quad (2)$$

$$[\mathrm{Co^{II}TsPc}]^{4-} \rightarrow [\mathrm{Co^{III}TsPc}]^{3-} + \mathrm{e}^-(E_{\mathrm{pa3}}) \quad (3)$$

$$[\mathrm{Co^{II}TsPc}]^{4-} + \mathrm{e}^- \rightarrow [\mathrm{Co^{I}TsPc}]^{5-}(E_{\mathrm{pc2}}) \quad (4)$$

$$[\mathrm{TsPc}]^{5-} + \mathrm{e}^- \rightarrow [\mathrm{TsPc}]^{6-}(E_{\mathrm{pc1}}) \quad (5)$$

Interestingly, ITO-{PAH/$Co^{II}TsPc$}$_3$ electrode in 0.1 mol L^{-1} PBS catalyzed oxidation of cysteine to cystine in the concentration range of 1.0×10^{-4}–1.6×10^{-3} mol L^{-1} at 0.4 V (*vs* SCE). The mechanism of cysteine oxidation catalyzed by this nanostructured electrode is depicted schematically in Fig. 5.5. Additionally, the electrochemical behavior of $Co^{II}TsPc$ nanostructured electrode was more sensitive than those with bare ITO and ITO-$Co^{II}TsPc$ electrodes.

Phenolic compounds are found in most fruits and vegetables. These endogenous compounds show interesting properties such as antioxidant activity, enzymatic inhibition and free radical scavenging action [42]. In particular, catechol is a diphenol compound of interest in the food industry and has been involved in glial cell toxicity. Thus, there is a great appeal to produce novel sensors of higher stability and lower cost for catechol detection [43].

Alessio and co-workers have combined the properties of iron phthalocyanine (FePc) and phospholipid dimyristoyl phosphatidic acid (DMPA) to produce LB films for the detection of phenolic compounds [43]. For this purpose, initially mixed FePc plus DMPA in chloroform solution was carefully spread onto phosphate buffer subphase (0.1 mol L^{-1} with NaCl 0.1 mol L^{-1}), while solvent was removed by evaporation within 15 min. The LB monolayer was obtained by symmetrical compression at 10 mm min^{-1} then transferred to the solid substrate. The sequential repetition of this process allowed the deposition of LB multilayers. In order to understand the electrochemical behavior of DMPA/FePc nanocomposite, the cast films of DMPA and FePc were prepared and voltammograms recorded. The phospholipid DMPA did not show electrochemical process in the potential window of -1+1 V (*vs* SCE), however the FePc exhibited a reduction peak at -0.60 V attributed to the macrocycle ring [26]. For the nanocomposite {DMPA+FePc} its voltammogram showed a peak assigned to the FePc species shifted to the reductive region (-0.77), suggesting an influence of DMPA on the electrochemical process. The LB film $\{DMPA + FePc\}_{10}$ immobilized onto ITO electrode was tested towards catechol and compared with bare ITO electrode. The anodic peak centered at 0.90 V for catechol group in the presence of bare ITO electrode shifted to 0.22 V when LB monolayers were incorporated. This data indicated an intense electrocatalytical performance probably associated with synergistic supramolecular effects between DMPA and FePc species. Moreover, LB film exhibited sensitivity and detection limit in the presence of catechol of 1.21 μM^{-1} and 0.43 μM, respectively. This LB platform is also promising for enzyme immobilization when a friendly environment is required [1].

Combining biological components (e.g., enzyme or DNA) with nanomaterials is a fast expanding research field that aims to develop novel nanostructure-based electrochemical biosensors. In order to achieve this goal, electroactive nanostructured membranes (ENM) were initially prepared through chemical immobilization of polyamidoamine dendrimer (PAMAM), gold nanoparticles (AuNPs) and polyvinylsulfonate (PVP) onto ITO substrate [44]. Electrocatalytical activity for H_2O_2 reduction by ENM containing several metal hexacyanoferrates (Ni, Fe, Cu and Co) was evaluated by voltammetric and impedance spectroscopy. All hexacyanoferrate-modified electrodes showed efficient H_2O_2 reaction, however, significant differences were observed in the electrochemical behavior. This behavior can probably be associated to the intimate contact between the redox mediator (rMe) and AuNPs, which improved the charge transfer within ENM as suggested by electrochemical impedance spectra. For example, nickel and copper hexacyanoferrates showed a decrease in anodic and cathodic peaks upon addition of 1.0×10^{-3} mol L^{-1} H_2O_2 and potential reduction at 0.2 V. For iron and cobalt

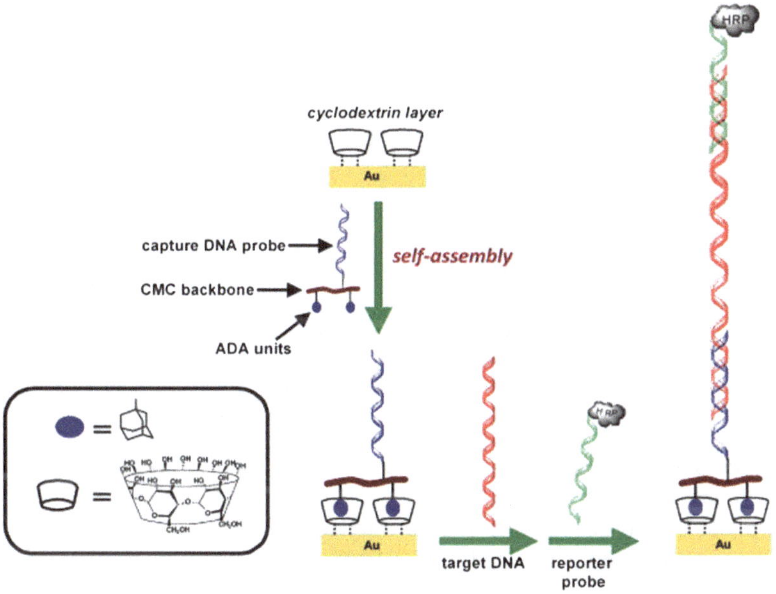

Fig. 5.6 Scheme adopted for the construction of the supramolecular genosensor. Carboxymethylcellulose (CMC) was used in order to modify the ADA surface and capture DNA probe. Reproduced with permission from [45]

hexacyanoferrates, an increase in the cathodic current at +0.1 V and a catalytic effect only at high H_2O_2 concentrations were observed. Based on this study, it is clear that the development of novel biosensors must involve both appropriate nanomaterials and self-assembly techniques [1].

In addition of building protein-based biosensors, DNA sensors can also be prepared onto modified surfaces using many immobilization approaches such as covalent binding and adsorption of specific oligonucleotide sequences among others [1]. Recently, Ortiz and co-workers [45] reported a novel strategy for the construction of supramolecular genosensors based on the interfacial self-assembly of bi-functionalized polymers bearing adamantine (ADA) and DNA onto a CD polymer surface, as illustrated in Fig. 5.6. In fact, the genosensor platform showed a linear response until 2 nmol L^{-1} exhibiting a sensitivity performance up to 0.35 nmol $L^{-1}/\mu A$ and lower limit of 80 pmol L^{-1} for the detection of a human leukocyte antigen allele associated with celiac disease.

Aptamer recognition has been described for the development of biosensors and applied to biomedical and environmental studies [46]. As illustrated in Fig. 5.7, an electrochemical aptamer-based sensor (E-AB) for specific recognition of thrombin was constructed. Zhang and co-workers [46] used immobilization of a Fe_3O_4-nanoparticles/tagged aptamer via a self-assembly method. In this case, bifunctional aptamer was covalently linked to both Fe_3O_4-NPs and gold electrode. Certainly

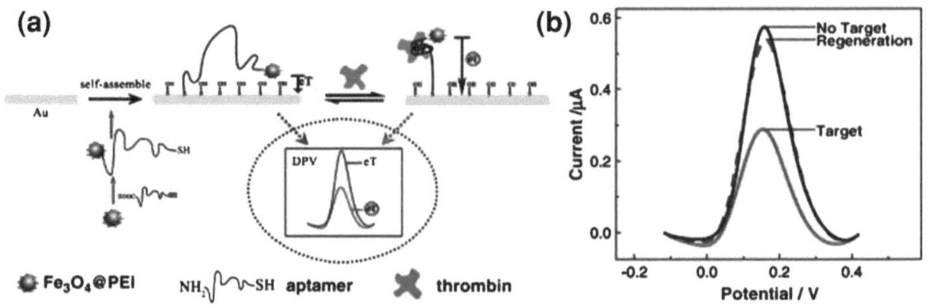

Fig. 5.7 **a** Schematic representation and thrombin recognition by an electrochemical aptamer-based sensor and **b** voltammogram curves of the E-AB sensor before and after deposition of 40 nM of thrombin. Reproduced with permission from [46]

electrochemical aptasensor could be used for clinical diagnosis, where fast responses and low cost are required.

Fabre and co-workers have reported interesting applications involving biomolecular recognition and controlled drug release using a quite simple strategy. They had prepared a chemically modified gold electrode with a monolayer that promoted hydrogen bonding interactions between adenine-substituted ferrocene and an uracil-terminated species showing a high electrochemical stability [47].

5.3 Nanobiological Sensors as a Natural Inspiration

"High-tech nanosensors" have been produced and improved by nature during millions of years of evolution. From single-celled species to human being these sensors have exhibited key physiological roles in striving adaptation and regulating life as we know [48–51]. Most of these sensors are protein-based macromolecules but RNA-based sensors known as riboswitches are also widespread and many recent discoveries have fueled this fast expanding field [52, 53]. Among these sensors, heme-based proteins are a quite remarkable example of versatility. Despite just recently hemeproteins were discovered to have a third function as a sensor, nowadays these proteins have been found in all kingdom of nature from archaea to human [51]. High degree of modularity has been noticed where many different heme folds (e.g., PAS, HNOB, CooA, GAF, Globin, SCHIC) are coupled to another variety of output protein domains [51, 54–56]. These heme-domains regulate a response promoted by the output domains, which usually carried out an enzymatic process, protein–protein or DNA–protein interactions. These nanomolecules are involved in sensing and responding to specific levels of gaseous biological molecules such as O_2, CO and NO (Fig. 5.8). They are able to distinct these diatomic molecules and report their levels in an impressive range of concentration and selective mechanisms as detailed elsewhere [51]. One aspect to

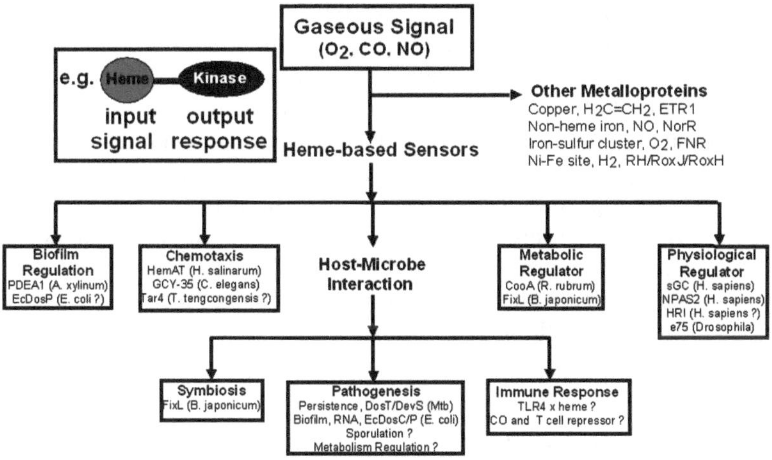

Fig. 5.8 Schcme for heme-based sensors and their biological roles

remark is the right usage of the term sensor that implies it must bind reversibly to signaling molecule. Thus, it can monitor continually ligand levels by reversible interactions, instead of detecting a biological signal by an irreversible process.

One interesting example of heme-based sensors can be found in Rhizobium bacteria living in symbioses that fix nitrogen gas for leguminous. Plant nodule roots must provide a suitable environment for bacteria to fix nitrogen where low levels of oxygen are an essential feature. This is required due to the extreme oxygen sensitivity of nitrogen fixation apparatus. If oxygen is present all efforts to prepare highly sophisticated proteins to convert nitrogen in ammonium will be wasted [57]. So, to coordinate this process it is essential to have an oxygen sensor for this duty. FixL was first identified as such sensor and has become a prototype heme-based sensor [57, 58]. Nowadays, many mechanistic details have emerged for FixL along with X-ray structures [51, 59–61]. This protein contains two main modules one heme-containing domain and a kinase domain (enzymatic). It works by transferring a phosphoryl group from ATP to another protein called FixJ. This latter phosphorylated works as a transcription factor inducing the expression of genes leading to produce a set of proteins responsible for nitrogen fixation and survival under microaerobic environment [51, 57]. Oxygen binds reversibly to FixL shutting off kinase activity so preventing production of many proteins but after a drop of oxygen levels it turns on this same system [51]. More recently, other similar systems have been identified in Mycobacterium tuberculosis (Mtb). DevS and DosT are also another oxygen heme-based sensor but involved in leading Mtb to a dormant or persistent state [54]. Persistent Mtb is very difficult to eliminate and might be responsible for the long time of tuberculosis treatment [62]. Designing inhibitors for these sensors are interesting strategies to shorten Mtb treatment and provide alternative drug targets and elaborate screening strategies are highly desired calling for nanotechnology assistance. Other sensors found in humans such as soluble guanylate cyclase (sGC) and NPAS2 are under intense investigation and involved in key events [63, 64]. The former is a biological sensor for nitric oxide (NO) involved in regulating important physiological process in vivo such as vasodilation, platelet aggregation and memory processes. NPAS2 is still under debate but might be involved in a biological human circadian clock with promising future applications. These systems tell us much about how biology has evolved to prepare highly well designed nanostructures to function as sensors and there is no question they are always a source of inspiration to nanotechnology field. Additionally, the increasing requirement to develop small-molecule regulators for these sensors have placed new exciting challenges for designing screening strategies to fast identify these target molecules.

5.4 Concluding Remarks

This chapter describes how the electrochemical modulation of sophisticated nanoplatforms, including sensors and biosensors, can be achieved taking advantage of the combination of: (i) supramolecular self-assembly method, (ii) hybrid nanomaterials, (iii) supramolecular chemistry concepts and (iv) electrochemical technique. It is expected in the close future that development of nanodevices with specific purpose will render more practicable if such approaches developed here find widespread use. Finally, we also explore the capability of proteins and enzymes for immobilization in the nanostructured systems aiming to construct nanobiological sensors with high control of electrochemical proprieties. Once more bioinspired, it is also expected we could take full advantage of protein modularity to build even more specific devices to manipulate kinase process or gene expression upon an specific signal, for example, or even to prepare nanomaterials for easier screening strategies to develop novel drugs.

Acknowledgments The authors gratefully acknowledge the financial support of the CAPES (nBioNet), CNPq (472369/2008-3 and 577410/2008-3 projects), FAPEPI, and PPP/FUNCAP.

References

1. Iost, R.M., Silva, W.C., Madurro, J.M., Brito-Madurro, A.G., Ferreira, L.F., Crespilho, F.N.: Recent advances in nano-based electrochemical biosensors: application in diagnosis and monitoring of diseases. Front. Biosci. **3**, 663–689 (2011). doi:10.2741/e278
2. Lehn, J.M.: From supramolecular chemistry towards constitutional dynamic chemistry and adaptive chemistry. Chem. Soc. Rev. **36**, 151–160 (2007). doi:10.1039/B616752G
3. Lehn, J.M.: Supramolecular chemistry: from molecular information towards self-organization and complex matter. Rep. Prog. Phys. **67**, 249–265 (2004). doi:10.1088/0034-4885/67/3/R02
4. D'Souza, F., Chitta, R., Sandanayaka, A.S.D., Subbaiyan, N.K., D'Souza, L., Araki, Y., Ito, O.: Supramolecular carbon nanotube-fullerene donor-acceptor hybrids for photoinduced electron transfer. J. Am. Chem. Soc. **129**, 15865–15871 (2007). doi:10.1021/ja073773x

5. Santos, A.C., Luz, R.A.S., Ferreira, L.G.F., Santos Júnior, J.R., Silva, W.C., Crespilho, F.N.: Organização supramolecular da ftalocianina de cobalto (II) e seu efeito na oxidação do aminoácido cisteína. Quim. Nova **33**, 539–546 (2010). doi:10.1590/S0100-40422010000300009
6. Alencar, W.S., Crespilho, F.N., Santos, M.R.M.C., Zucolotto, V., Oliveira Júnior, O.N., Silva, W.C.: Influence of film architecture on the charge-transfer reactions of metallophthalocyanine layer-by-layer films. J. Phys. Chem. C **111**, 12817–12821 (2007). doi:10.1021/jp070695r
7. Alencar, W.S., Crespilho, F.N., Martins, M.V.A., Zucolotto, V., Oliveira Júnior, O.N., Silva, W.C.: Synergistic interaction between gold nanoparticles and nickel phthalocyanine in layer-by-layer (LbL) films: evidence of constitutional dynamic chemistry (CDC). Phys. Chem. Chem. Phys. **11**, 5086–5091 (2009). doi:10.1039/b821915j
8. Luz, R.A.S., Martins, M.V.A., Magalhães, J.L., Siqueira Júnior, J.R., Zucolotto, V., Oliveira Júnior, O.N., Crespilho, F.N., Silva, W.C.: Supramolecular architectures in layer-by-layer films of single-walled carbon nanotubes, chitosan and cobalt (II) phthalocyanine. Mater. Chem. Phys. (2011). doi:10.1016/j.matchemphys.2011.08.038
9. Escosura-Muñiz, A., Merkoçi, A.: Electrochemical detection of proteins using nanoparticles: applications to diagnostics. Expert Opin. Med. Diagn. **4**, 21–37 (2010). doi:10.1517/17530050903386661
10. Yang, S., Xu, B., Zhang, J., Huang, X., Ye, J., Yu, C.: Controllable adsorption of reduced graphene oxide onto self-assembled alkanethiol monolayers on gold electrodes: tunable electrode dimension and potential electrochemical applications. J. Phys. Chem. C **114**, 4389–4393 (2010). doi:10.1021/jp911760b
11. Lehmann, P., Symietz, C., Brezesinski, G., Krass, H., Kurth, D.G.: Langmuir and langmuir-blodgett films of metallosupramolecular polyelectrolyte-amphiphile complexs. Langmuir **21**, 5901–5906 (2005). doi:10.1021/la050841p
12. Decher, G.: Fuzzy nanoassemblies: toward layered polymeric multicomposites. Science **277**, 1232–1237 (1997). doi:10.1126/science.277.5330.1232
13. Ariga, K., Hill, J.P., Ji, Q.: Layer-by-layer assembly as a versatile bottom-up nanofabrication technique for exploratory research and realistic application. Phys. Chem. Chem. Phys. **9**, 2319–2340 (2007). doi:10.1039/b700410a
14. Alessio, P., Rodríguez-Méndez, M.L., Saez, J.A.S., Constantino, C.J.L.: Iron phthalocyanine in non-aqueous medium forming layer-by-layer films: growth mechanism, molecular architecture and applications. Phys. Chem. Chem. Phys. **12**, 3972–3983 (2010). doi:10.1039/b922242c
15. Ferreyra, N., Coche-Guérente, L., Fatisson, J., Teijelo, M.L., Labbé, P.: Layer-by-layer self-assembled multilayers of redox polyelectrolytes and gold nanoparticles. Chem. Commun. 2056–2057 (2003). doi:10.1039/B305347D
16. Daniel, M.C., Astruc, D.: Gold Nanoparticles: assembly, supramolecular chemistry, quantum-size-related properties, and applications toward biology, catalysis, and nanotechnology. Chem. Rev. **104**, 293–346 (2004). doi:10.1021/cr030698+
17. Cathell, M.D., Szewczyk, J.C., Bui, F.A., Weber, C.A., Wolever, J.D., Kang, J., Schauer, C.L.: Structurally colored thiol chitosan thin films as a platform for aqueous heavy metal ion detection. Biomacromolecules **9**, 289–295 (2008). doi:10.1021/bm700845z
18. Li, X., Zhou, Y., Zheng, Z., Yue, X., Dai, Z., Liu, S., Tang, Z.: Glucose biosensor based on nanocomposite films of CdTe quantum dots and glucose oxidase. Langmuir **25**, 6580–6586 (2009). doi:10.1021/la900066z
19. Alarcón-Ángeles, G., Guix, M., Silva, W.C., Ramírez-Silva, M.T., Palomar-Pardavé, M., Romero-Romo, M., Merkoçi, A.: Enzyme entrapment by β-cyclodextrin electropolymerization onto a carbon nanotubes-modified screen-printed electrode. Biosens. Bioelectron. **26**, 1768–1773 (2010). doi:10.1016/j.bios.2010.08.058
20. Mani, V., Chikkaveeraiah, B.V., Patel, V., Gutkind, J.S., Rusling, J.F.: Ultrasensitive immunosensor for cancer biomarker proteins using gold nanoparticle film electrodes and multienzyme-particle amplification. ACS Nano **3**, 585–594 (2009). doi:10.1021/nn800863w

21. Fan, H., Xing, R., Xu, Y., Wang, Q., He, P., Fang, Y.: A new electrochemical method for DNA sequence detection with homogeneous hybridization based on host–guest recognition technology. Electrochem. Commun. **12**, 501–504 (2010). doi:10.1016/j.elecom.2009.11.030
22. Zucolotto, V., Daghastanli, K.R.P., Hayasaka, C.O., Riul Júnior, A., Ciancaglini, P., Oliveira Júnior, O.N.: Using capacitance measurements as the detection method in antigen-containing layer-by-layer films for biosensing. Anal. Chem. 79, 2163–2167 (2007). doi:10.1021/ac0616153
23. Willner, I., Willner, B.: Biomolecule-based nanomaterials and nanostructures. Nano Lett. **10**, 3805–3815 (2010). doi:10.1021/nl102083j
24. Feldheim, D.L., Eaton, B.E.: Selection of biomolecules capable of mediating the formation of nanocrystals. ACS Nano **1**, 154–159 (2007). doi:10.1021/nn7002019
25. Ariga, K., Hill, J.P., Lee, M.V., Vinu, A., Charvet, R., Acharya, S.: Challenges and breakthroughs in recent research on self-assembly. Sci. Technol. Adv. Mater. **9**, 1–96 (2008). doi:10.1088/1468-6996/9/1/014109
26. Crespilho, F.N., Silva, W.C., Zucolotto, V.: Supramolecular assemblies of metallophthalocyanines: physicochemical properties and applications. In: Nantes, I.L. (ed.) Catalysis and Photochemistry in Heterogeneous Media, 1st edn. Research Signpost, Kerala (2007)
27. Nijhuis, C.A., Dolatowska, K.A., Ravoo, B.J., Huskens, J., Reinhoudt, D.N.: Redox-controlled interaction of biferrocenyl-terminated dendrimers with β-cyclodextrin molecular printboards. Chem. Eur. J. **13**, 69–80 (2007). doi:10.1002/chem.200600777
28. Schmidt, D.J., Cebeci, F.C., Kalcioglu, Z.I., Wyman, S.G., Ortiz, C., Vliet, K.J.V., Hammond, P.T.: Electrochemically controlled swelling and mechanical properties of a polymer nanocomposite. ACS Nano **3**, 2207–2216 (2009). doi:10.1021/nn900526c
29. Zhao, W., Xu, J.J., Shi, C.G., Chen, H.Y.: Multilayer membranes via layer-by-layer deposition of organic polymer protected prussian blue nanoparticles and glucose oxidase for glucose biosensing. Langmuir **21**, 9630–9634 (2005). doi:10.1021/la051370+
30. Silva, W.C., Guix, M., Angeles, G.A., Merkoçi, A.: Compact microcubic structures platform based on self-assembly prussian blue nanoparticles with highly tuneable conductivity. Phys. Chem. Chem. Phys. **12**, 15505–15511 (2010). doi:10.1039/c0cp00960a
31. Siqueira Júnior, J.R., Albouzar, M.H., Poghossian, A., Zucolotto, V., Oliveira Júnior, O.N., Schöning, M.J.: Penicillin biosensor based on a capacitive field-effect structure functionalized with a dendrimer/carbon nanotube multilayer. Biosens. Bioelectron. **25**, 497–501 (2009). doi:10.1016/j.bios.2009.07.007
32. Kim, S.N., Rusling, J.F., Papadimitrakopoulos, F.: Carbon nanotubes for electronic and electrochemical detection of biomolecules. Adv. Mater. **19**, 3214–3228 (2007). doi:10.1002/adma.200700665
33. Rivas, G.A., Rubianes, M.D., Rodríguez, M.C., Ferreyra, N.F., Luque, G.L., Pedano, M.L., Miscoria, A.S., Parrado, C.: Carbon nanotubes for electrochemical biosensing. Talanta **74**, 291–307 (2007). doi:10.1016/j.talanta.2007.10.013
34. Sun, L., Zhang, L., Liang, C., Yuan, Z., Zhang, Y., Xu, W., Zhang, J., Chen, Y.: Chitosan modified Fe^0 nanowires in porous anodic alumina and their application for the removal of hexavalent chromium from water. J. Mater. Chem. **21**, 5877–5880 (2011). doi:10.1039/c1jm10205b
35. Jayakumar, R., Menon, D., Manzoor, K., Nair, S.V., Tamura, H.: Biomedical applications of chitin and chitosan based nanomaterials-a short review. Carbohydr. Polym. **82**, 227–232 (2010). doi:10.1016/j.carbpol.2010.04.074
36. Li, D., Kaner, R.B.: Graphene-based materials nanoelectronics and many other applications. Science **320**, 1170–1171 (2008). doi:10.1126/science.1158180
37. Allen, M.J., Tung, V.C., Kaner, R.B.: Honeycomb carbon: a review of graphene. Chem. Rev. **110**, 132–145 (2010). doi:10.1021/cr900070d
38. Shao, Y., Wang, J., Wu, H., Liu, J., Aksay, I.A., Lin, Y.: Graphene based electrochemical sensors and biosensors: a review. Electroanalysis **22**, 1027–1036 (2010). doi:10.1002/elan.200900571

39. Li, M.J., Zhan, C.Q., Nie, M.J., Chen, G.N., Chen, X.: Selective recognition of homocysteine and cysteine based on new ruthenium(II) complexes. J. Inorg. Biochem. **105**, 420–425 (2011). doi:10.1016/j.jinorgbio.2010.12.007
40. Sekota, M., Nyokong, T.: The study of the interactions of cobalt(II) tetrasulfophthalocyanine with cysteine and histidine. Polyhedron **16**, 3279–3284 (1997). doi:10.1016/S0277-5387(97)00096-X
41. Luz, R.C.S., Moreira, A.B., Damos, F.S., Tanaka, A.A., Kubota, L.T.: Cobalt tetrasulphonated phthalocyanine immobilized on poly-L-lysine film onto glassy carbon electrode as amperometric sensor for cysteine. J. Pharm. Biomed. Anal. **42**, 184–191 (2006). doi:10.1016/j.jpba.2006.03.036
42. Dewick, P.M.: Medicinal Natural Products: A Biosynthetic Approach. Wiley, New York (2002)
43. Alessio, P., Pavinatto, F.J., Oliveira Júnior, O.N., Saez, J.A.S., Constantino, C.J.L., Rodríguez-Méndez, M.L.: Detection of catechol using mixed Langmuir–Blodgett films of a phospholipid and phthalocyanines as voltammetric sensors. Analyst **135**, 2591–1599 (2010). doi:10.1039/c0an00159g
44. Crespilho, F.N., Ghica, M.E., Zucolotto, V., Nart, F.C., Oliveira Júnior, O.N., Brett, C.M.A.: Electroactive nanostructured membranes (ENM): synthesis and electrochemical properties of redox mediator-modified gold nanoparticles using a dendrimer layer-by-layer approach. Electroanalysis **19**, 805–812 (2007). doi:10.1002/elan.200603775
45. Ortiz, M., Torréns, M., Alakulppi, N., Strömbom, L., Fragoso, A., O'Sullivan, C.K.: Amperometric supramolecular genosensor self-assembled on cyclodextrin-modified surfaces. Electrochem. Commun. **13**, 578–581 (2011). doi:10.1016/j.elecom.2011.03.014
46. Zhang, S., Zhou, G., Xu, X., Cao, L., Liang, G., Chen, H., Liu, B., Kong, J.: Development of an electrochemical aptamer-based sensor with a sensitive Fe_3O_4 nanopaticle-redox tag for reagentless protein detection. Electrochem. Commun. **13**, 928–931 (2011). doi:10.1016/j.elecom.2011.06.002
47. Fabre, B., Ababou-Girard, S., Singh, P., Kumar, J., Verma, S., Bianco, A.: Noncovalent assembly of ferrocene on modified gold surfaces mediated by uracil–adenine base pairs. Electrochem. Commun. **12**, 831–834 (2010). doi:10.1016/j.elecom.2010.03.045
48. DeMaria, S., Ngai, J.: The cell biology of smell. J. Cell Biol. **191**, 443–452 (2010). doi:10.1083/jcb.201008163
49. Pedra, J.H., Cassel, S.L., Sutterwala, F.S.: Sensing pathogens and danger signals by the inflammasome. Curr. Opin. Immunol. **21**, 10–16 (2009). doi:10.1016/j.coi.2009.01.006
50. Giedroc, D.P., Arunkumar, A.I.: Metal sensor proteins: nature's metalloregulated allosteric switches. Dalton Trans. **7**, 3107–3120 (2007). doi:10.1039/B706769K
51. Gilles-Gonzalez, M.A., Gonzalez, G.: Heme-based sensors: defining characteristics, recent developments, and regulatory hypotheses. J. Inorg. Biochem. **99**, 1–22 (2005). doi:10.1016/j.jinorgbio.2004.11.006
52. Dambach, M.D., Winkler, W.C.: Expanding roles for metabolite-sensing regulatory RNAs. Curr. Opin. Microbiol. **12**, 161–169 (2009). doi:10.1016/j.mib.2009.01.012
53. Tucker, B.J., Breaker, R.R.: Riboswitches as versatile gene control elements. Curr. Opin. Struct. Biol. **15**, 342–348 (2005). doi:10.1016/j.sbi.2005.05.003
54. Sousa, E.H.S., Tuckerman, J.R., Gonzalez, G., Gilles-Gonzalez, M.A.: DosT and DevS are oxygen-switched kinases in Mycobacterium tuberculosis. Protein Sci. **16**, 1708–1719 (2007). doi:10.1110/ps.072897707
55. Moskvin, O.V., Kaplan, S., Gilles-Gonzalez, M.A., Gomelsky, M.: Novel heme-based oxygen sensor with a revealing evolutionary history. J. Biol. Chem. **282**, 28740–28748 (2007). doi:10.1074/jbc.M703261200
56. Reinking, J., Lam, M.M.S., Pardee, K., Sampson, H.M., Liu, S., Yang, P., Williams, S., White, W., Lajoie, G., Edwards, A., Krause, H.M.: The drosophila nuclear receptor e75 contains heme and is gas responsive. Cell **122**, 195–207 (2005). doi:10.1016/j.cell.2005.07.005

57. Gilles-Gonzalez, M.A., Gonzalez, G.: Signal transduction by heme-containing PAS-domain proteins. J. Appl. Physiol. **96**, 774–783 (2004). doi:10.1152/japplphysiol.00941.2003
58. Gilles-Gonzalez, M.A., Ditta, G.S., Helinski, D.R.: A haemoprotein with kinase activity encoded by the oxygen sensor of Rhizobium meliloti. Nature **350**, 170–172 (1991). doi:10.1038/35017a0
59. Sousa, E.H.S., Tuckerman, J.R., Gonzalez, G., Gilles-Gonzalez, M.A.: A memory of oxygen binding explains the dose response of the heme-based sensor FixL. Biochemistry **46**, 6249–6257 (2007). doi:10.1021/bi7003334
60. Gilles-Gonzalez, M.A., Caceres, A.I., Sousa, E.H.S., Tomchick, D.R., Brautigam, C.A., Gonzalez, C., Machius, M.: A proximal arginine R206 participates in switching of the Bradyrhizobium japonicum FixL oxygen sensor. J. Mol. Biol. **360**, 80–89 (2006). doi:10.1016/j.jmb.2006.04.054
61. Sousa, E.H.S., Gonzalez, G., Gilles-Gonzalez, M.A.: Oxygen blocks the reaction of the FixL-FixJ complex with ATP but does not influence binding of FixJ or ATP to FixL. Biochemistry **44**, 15359–15365 (2005). doi:10.1021/bi051661h
62. Sacchettini, J.C., Rubin, E.J., Freundlich, J.S.: Drugs versus bugs: in pursuit of the persistent predator mycobacterium tuberculosis. Nat. Rev. Microbiol. **6**, 41–52 (2008). doi:10.1038/nrmicro1816
63. Sousa, E.H.S., Gonzalez, G., Gilles-Gonzalez, M.A.: Soluble guanylylcyclase and its microbial relatives. In: Ghosh, A. (ed.) The Smallest Biomolecules: Diatomics and their Interactions with Heme Proteins, 1st edn. Elsevier, Amsterdam (2008)
64. Dioum, E.M., Rutter, J., Tuckerman, J.R., Gonzalez, G., Gilles-Gonzalez, M.A., McKnight, S.L.: NPAS2: a gas-responsive transcription factor. Science **298**, 2385–3287 (2002). doi:10.1126/science.1078456

Chapter 6
DNA and Enzyme-Based Electrochemical Biosensors: Electrochemistry and AFM Surface Characterization

Christopher Brett and Ana Maria Oliveira-Brett

Abstract The characterization and applications of nanofilms on biologically modified electrode surface processes opens up exciting new prospects for designing new forms of matter. This chapter will summarise and illustrate recent developments on surface characterisation of DNA and enzyme-based sensors to complement information obtained by electrochemical and impedance techniques. The DNA-electrochemical biosensor incorporates immobilised DNA as molecular recognition element on the electrode surface, and measures specific binding processes with DNA, enabling the screening and evaluation of the effect caused to DNA by health hazardous compounds and oxidising substances. AFM imaging is used to characterize different procedures for immobilising nanoscale double-stranded DNA surface films on carbon electrodes, in which a critical issue is the sensor material and the degree of surface coverage. The DNA-electrochemical biosensor gives very important mechanistic information because the mechanisms of DNA-hazard compound interaction at charged interfaces mimic the in vivo situation. Electrochemical enzyme biosensors consist of electrodes modified with one or more layers containing the immobilized enzyme, and possibly a redox mediator. Operation depends very much on the surface exposed to solution, and the nanostructure conditions the access of analyte to the enzyme active sites and its electroactive products to the electrode substrate. Full characterization of the assembly, both morphological, as well as structural and electrical—by electrochemical voltammetric and impedance techniques—is thus crucial to effective

C. Brett (✉) · A. M. Oliveira-Brett (✉)
Departamento de Química, Faculdade de Ciências e Tecnologia, Universidade de Coimbra, 3004-535, Coimbra, Portugal
e-mail: brett@ci.uc.pt

F. N. Crespilho (ed.), *Nanobioelectrochemistry*, DOI: 10.1007/978-3-642-29250-7_6,

nanostructuring of the enzyme sensor electrode. These questions will be surveyed and discussed in the light of recent research and some future directions will be indicated.

6.1 Introduction

The characterization and applications of nanofilms on biologically modified electrode surface processes opens up exciting new prospects for designing new forms of matter. The search for efficient, rapid-response electrochemical biosensors has led to the development of new strategies for their construction and to the search for better materials, bearing in mind the requirements of fast electrode kinetics, fast mass transport of analyte species and sensor simplicity. In this context, the importance of microsystems, information acquisition and use, new materials, and sensor operation can be identified [1]. Most of these questions are intimately linked to the general umbrella of "nanotechnology". Materials important for biosensors include nanostructured biomaterials, metals and alloys, different forms of carbon, electroactive and conducting polymers.

This chapter will illustrate recent developments on surface characterisation of DNA and enzyme-based biosensors to complement information obtained by electrochemical and impedance techniques.

In recent years increased attention has been focused on the ways in which hazard compounds and anti-cancer drugs interact with DNA, with the goal of understanding the toxic as well as chemotherapeutic effects of many molecules. The development of fast and accurate methods of oxidative DNA damage detection is important.

The DNA-electrochemical biosensor is a very good model for evaluation of nucleic acid damage, and electrochemical detection is a particularly sensitive and selective method for the investigation of specific interactions [2–6]. The interpretation of electrochemical data can contribute to elucidation of the mechanism by which DNA is oxidatively damaged by hazardous compounds, in an approach to the real action scenario that occurs in the living cell and without using animal tests.

Some recent developments in materials for use in electrochemical enzyme biosensors will illustrate the strategy of sensor build-up and sensor characterisation by electrochemical and non-electrochemical techniques, illustrated by the type of information that has been obtained at the molecular and nanometre levels. Any construction strategy that is developed has to consider easy access to the enzyme active site by the enzyme substrate and easy removal of products with a convenient transduction mechanism for production of an electrical signal. Characterisation of different approaches will be presented.

6.2 DNA-Electrochemical Biosensors

The electrochemical sensor for detecting DNA damage consists of a glassy carbon electrode with DNA immobilized on its surface. The possibility of foreseeing the damage that hazard compounds cause to DNA integrity arises from the pre-concentration of either the starting materials or the redox reaction products on the DNA-biosensor surface, thus enabling electrochemical probing of the presence of short-lived radical intermediates and of their damage to dsDNA.

AFM images were used to characterize different procedures for immobilization of nanoscale DNA surface films on carbon electrodes before and after interaction with hazard compounds. In the development and design of DNA-electrochemical biosensors it is very important to know the DNA structure, the variations in DNA conformations—polymorphisms—and to understand the electrochemical behaviour of DNA molecules on the electrochemical transducer.

The electrochemical transduction is dynamic in that the electrode is itself a tuneable charged reagent as well as a detector of all surface phenomena, which greatly enlarges the electrochemical biosensing capabilities.

The development and characterization of a DNA-electrochemical biosensor provides very relevant information because the mechanisms of DNA-hazard compound interaction at charged interfaces mimic better the in vivo situation. The detection of chemical compounds that cause irreversible damage to DNA is very important, as they may lead to hereditary or carcinogenic diseases. Reactions with chemical substances cause changes in the structure of DNA and the base sequence leading DNA oxidative damage and to perturbations in DNA replication.

Electrode surface modification has been done by different DNA adsorption immobilization procedures, electrostatic adsorption or evaporation, with the formation of a monolayer or a multilayer DNA film. A very important factor for the optimal construction of a DNA-electrochemical biosensor is the immobilization of the DNA probe on the electrode surface [2–6].

There are different procedures that can be followed in the DNA-electrochemical biosensor construction depending on the required application [7–9].

6.2.1 AFM Surface Characterization

A critical issue in the development of an electrochemical DNA-biosensor is the sensor material and the degree of surface coverage. MAC Mode AFM images were used to characterize different procedures for immobilising nanoscale double-stranded DNA (ds-DNA) surface nanofilms on carbon electrodes, Fig. 6.1.

The results demonstrated that the hydrophobic interactions with the HOPG surface, Fig. 6.1a, explain the main adsorption mechanism, although other effects such as electrostatic and Van der Waals interactions may contribute to the adsorption process.

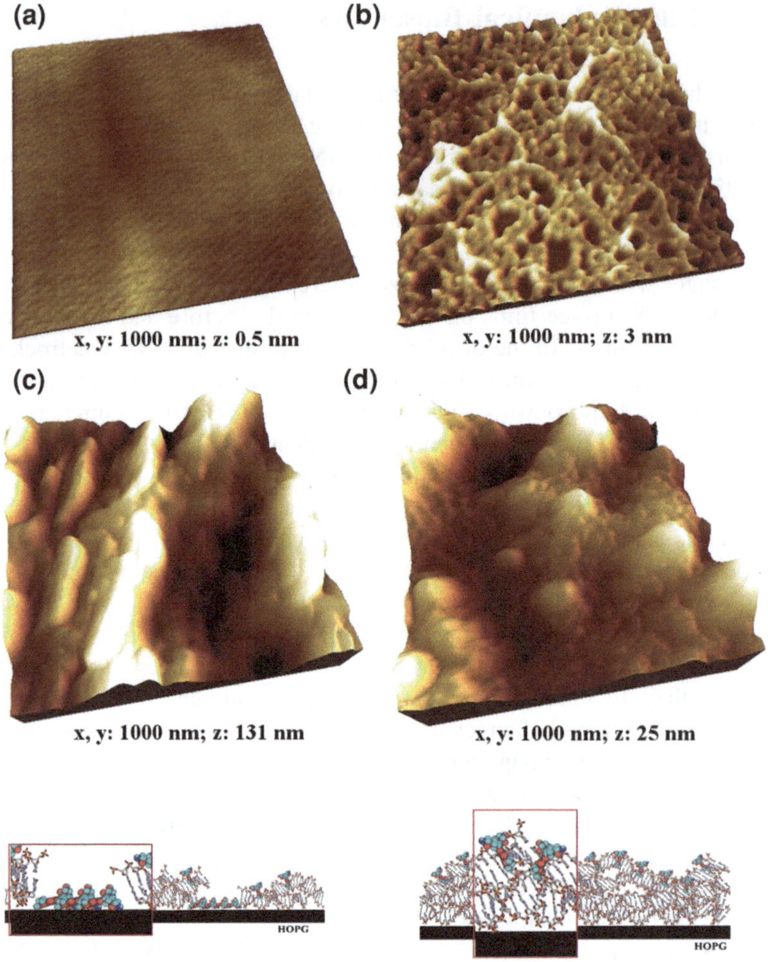

Fig. 6.1 MAC Mode AFM three-dimensional images in air of: **a** clean HOPG electrode; **b** thin layer dsDNA-biosensor surface, prepared onto HOPG by 3 min free adsorption from 60 μg/mL ds-DNA in pH 4.5 0.1 M acetate buffer; **c** multi-layer film dsDNA- electrochemical biosensor, prepared onto HOPG by evaporation of 3 consecutive drops each containing 5 μL of 50 μg/mL dsDNA in pH 4.5 0.1 M acetate buffer; **d** thick layer dsDNA-electrochemical biosensor, prepared onto HOPG by evaporation from 37.5 mg/mL dsDNA in pH 4.5 0.1 M acetate buffer; **e**, **f** Schematic models of dsDNA-anticancer drugs interaction using **e** the thin layer and **f** the thick layer dsDNA-electrochemical biosensor. [From Ref. [5] with permission]

Three procedures used in the DNA-electrochemical biosensor preparation by adsorption with or without applied potential, and their characterisation by AFM, are shown in Fig. 6.1:

1. *Thin-layer dsDNA biosensor*: prepared by immersing the GCE (d = 1.5 mm) surface in a 60 μg mL^{-1} dsDNA solution at +0.30 V applied potential during 10 min, Fig. 6.1b.
2. *Multi-layer dsDNA biosensor*: prepared by successively covering the GCE (d = 1.5 mm) surface with three drops of 5 μL each of 50 μg mL^{-1} dsDNA solution. After placing each drop on the electrode surface the biosensor is dried under a constant flux of N_2, Fig. 6.1c.
3. *Thick-layer dsDNA biosensor*: prepared by covering the GCE (d = 1.5 mm) surface with 10 μL of 35 mg mL^{-1} dsDNA solution and allowing it to dry in normal atmosphere, Fig. 6.1d.

The thin dsDNA nanolayer does not completely cover the HOPG electrode surface and the network structure has holes exposing the electrode underneath. The AFM image of a thin-layer dsDNA-electrochemical biosensor prepared on HOPG substrate is given in Fig. 6.1b, showing that the dsDNA molecules adsorbed on the HOPG surface form a two-dimensional lattice with uniform coverage of the electrode.

The DNA network patterns define nanoelectrode systems with different active surface areas on the graphite substrate, and form a nanobiomaterial matrix to attach and study interactions with hazard molecules. Hazard molecules from the bulk solution will also diffuse and adsorb non-specifically on the electrode's uncovered regions.

Both the multi- and thick-layer dsDNA-electrochemical biosensor preparation give rise to complete coverage of the electrode surface, with regularly dispersed peaks and valleys as shown by AFM images, Fig. 6.1c and d.

The dsDNA-electrode surface interactions are stronger and these DNA layers are very stable on the HOPG surface. The advantage of the multi-layer dsDNA biosensor with respect to the thick-layer is the short time necessary for the multi-layer ds-DNA-electrochemical biosensor construction.

6.2.2 Electrochemistry

DNA damage is caused by hazard compounds or their metabolites and leads to multiple modifications in DNA, including DNA strand breaks, base-free sites and oxidized bases. Oxidative DNA damage caused for instance by oxygen-free radicals leads to multiple modifications in DNA, including base-free sites and oxidised bases that are potentially mutagenic [10, 11].

The dsDNA biosensor has been used to study the influence of reactive oxygen species (ROS) in the mechanism of DNA damage, disrupting the helix and causing the formation of the biomarker 8-oxo-7,8-dihydro-2′-deoxyguanosine (8-oxodGuo), by a number of substances, such as neoplasic drugs, thalidomide, palladium complexes, and antioxidants.

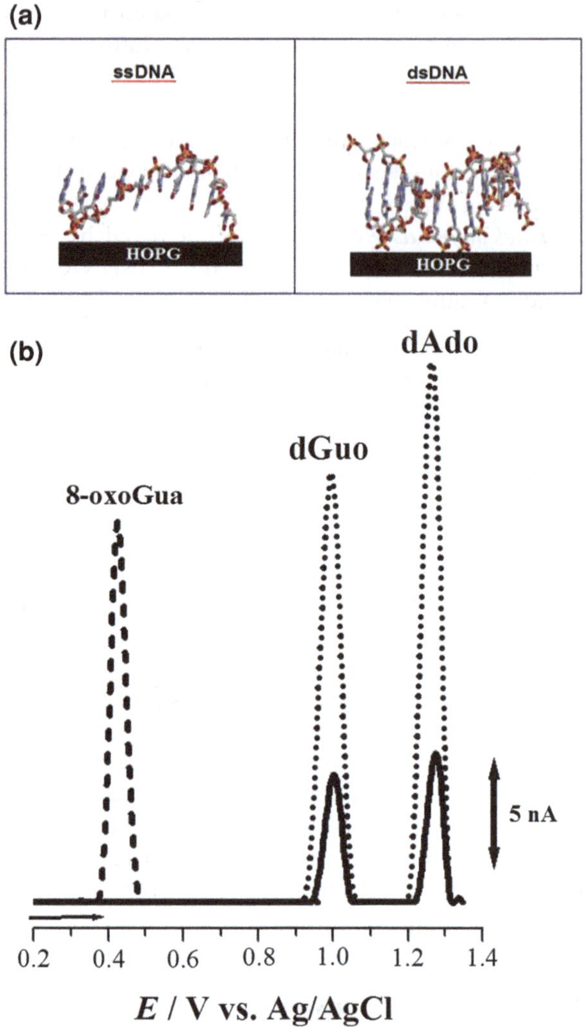

Fig. 6.2 **a** Schematic representation of ssDNA and dsDNA immobilized on the carbon electrode and **b** differential pulse voltammograms baseline corrected of (- - -) 5 μM 8-oxoGua or 2,8-oxoAde, (···) 60 μg/mL ssDNA and (—) 60 μg/mL dsDNA in pH 4.5 0.1 M acetate buffer. Pulse amplitude 50 mV, pulse width 70 ms, scan rate 5 mV/s. [From Ref. [33] with permission]

The major products of DNA oxidative damage are 8-oxo-7,8-dihydroguanine (8-oxoguanine, 8-oxoGua) which is the product of oxidation of guanine, the most easily oxidised base in DNA, and 2,8-dihydroxyadenine (2,8-oxoAde) [12], Fig. 6.2, and they can cause mutagenesis, hence the importance of screening for their occurrence.

The 8-oxoguanine mutagenicity causes loss of DNA base pairing specificity [12–14], and has been the subject of intensive research becoming widely accepted as a biomarker of oxidative DNA damage and cellular oxidative stress [15, 16]. Oxidative stress in vivo is an imbalance between prooxidant and antioxidant reactions which causes disruption of the redox mechanisms.

Elevated levels of 8-oxodGua were found in the urine and lung tissues of smokers [17] as well as in body fluids and DNA from human tissues of patients with disorders such as cancer, atherosclerosis, chronic hepatitis, cystic fibrosis, diabetes, acquired immunodeficiency syndrome, neurodegenerative and age-related diseases [18].

The electroactivity of 8-oxoGua and of 2,8-oxoAde was investigated by voltammetry [19, 20] and their oxidation peaks were chosen to monitor DNA oxidative damage since the oxidation potentials are lower than the oxidation potentials of the respective bases, they are easily detectable electrochemically and there is no overlap with the electrochemical oxidation peaks related to DNA.

The electrochemical detection of these biomarkers has been essential for investigating the electrochemical mechanisms of DNA oxidative damage using the DNA-electrochemical biosensor [21–39].

The approach described can be used for the understanding of ds-DNA interactions with various complex agents and individual chemicals of environmental, food and medical interest. The use of voltammetric inexpensive and fast detection techniques for the in situ generation of reactive intermediates is, in a successful way, a complementary tool for the study of biomolecular interaction mechanisms.

6.2.3 Applications of DNA-Electrochemical Biosensors

The investigation of DNA-drug interactions and the development of a fast and accurate electrochemical method for the detection of DNA oxidative damage, Fig. 6.3, especially from anticancer drugs used in cancer therapy or in the development of new antineoplasic drugs, has been growing in recent years because DNA-electrochemical biosensors are a very good model for simulating nucleic acid interactions and to clarify mechanisms of action.

The interaction with dsDNA of several anticancer drugs [21–39], namely, adriamycin, imatinib, thalidomide, Fig. 6.3, palladium compounds, and nucleoside analogues was investigated using the dsDNA-electrochemical biosensor or the interaction with dsDNA evaluated directly in the solution by electrochemical techniques.

The damage to immobilised DNA causes the appearance of oxidation peaks from DNA guanine and adenine purine bases which should always be detected. In the case of occurrence of DNA oxidative damage, the oxidation peaks from the biomarkers for oxidative damage, 8-oxoGua or 2,8-oxoAde, should be evaluated.

Aptamers are nucleic acid sequences (DNA or RNA) selected in vitro from large combinatorial pools to bind to specific targets ranging from small molecules to biological molecules and even cells. Aptamers exhibit a strong and specific binding affinity towards their targets and can be simply synthesized via cost-effective and readily automated routes, possessing significant advantages over other recognition molecules, such as antibodies, due to their small size, chemical simplicity, and flexibility.

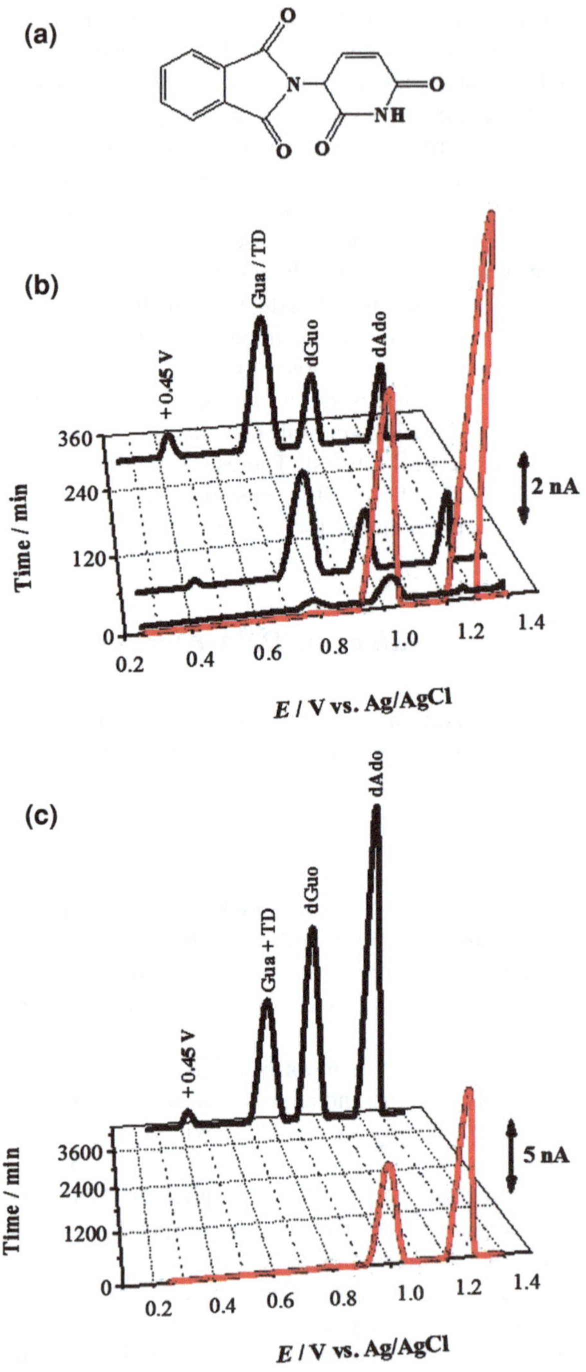

Fig. 6.3 **a** Chemical structure of thalidomide and **b**, **c** baseline corrected differential pulse voltammograms of (—) control 100 μg/mL dsDNA and (—) incubated solutions in pH 4.5 0.1 M acetate buffer of: **b** 100 μg/mL dsDNA with ~40 μM TD during 10 min, 1 h, 5 h and 24 h, and **c** 100 μg/mL dsDNA with ~40 μM TD during 72 h. [From Ref. [28] with permission]

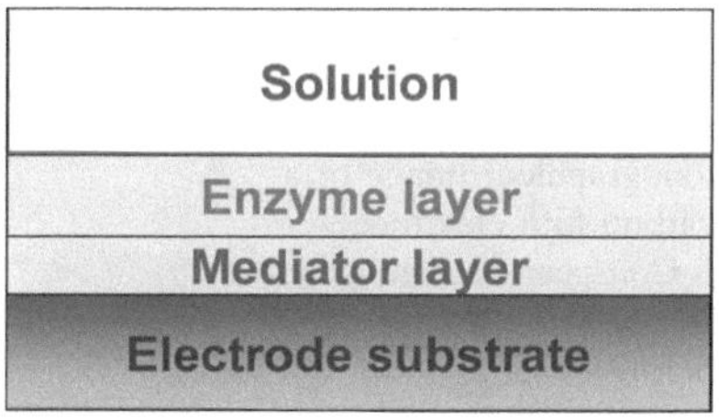

Fig. 6.4 Schematic construction of redox-mediated electrochemical enzyme biosensor

One of the most preeminent examples of the in vitro selection of DNA oligonucleotides for targeting a specific protein is the thrombin-binding aptamer (TBA). Thrombin is a serine protease and a coagulation protein in the blood stream that has many effects in the coagulation mechanism. Because of its importance in anticlotting therapeutics, TBA has been studied extensively and has also been used for the development of TBA-electrochemical biosensors for the detection of thrombin [40, 41].

The dsDNA-electrochemical biosensors are sophisticated devices to investigate hazard compounds and anticancer drugs-DNA interaction, and may contribute to drug discovery and effective treatment for cancer through providing knowledge about the efficacy of new drug binding with DNA and through information on the mechanism of drug-DNA interactions, and to the understanding of DNA interaction with molecules or ions and the mechanism through which DNA is oxidatively damaged without using animal tests.

6.3 Electrochemical Enzyme Biosensors

Considerable effort has been dedicated to the investigation of different surface modification processes of metal and carbon electrodes in the construction of electrochemical enzyme biosensors with separate redox mediator and enzyme layers, see Fig. 6.4.

The design must take into account the fact that a good redox mediator should have fast electrode kinetics, be stable and have a low redox potential in order to be easily regenerated in a biosensor context. Other modifications have concerned the use of carbon nanotubes, either immobilized by electrostatic adsorption or covalently immobilized in chitosan matrices and layer-by-layer self-assembly of sensing structures. Some of these strategies will be addressed in the following sections.

Recently a review has been published concerning the use of AFM, scanning electrochemical microscopy (SECM) and quartz crystal microbalance (QCM) as complementary techniques for the characterisation of enzyme-based bioanalytical platforms [42].

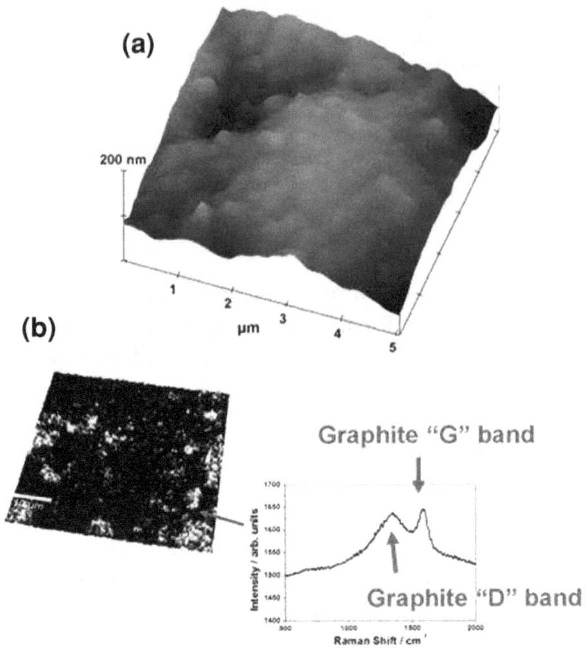

Fig. 6.5 Images of the surface of 1.5 Ω carbon film electrodes. **a** AFM topographical image of a carbon film electrode. **b** Confocal Raman spectrum. [From Ref. 44 with permission]

6.3.1 Carbon Electrode Substrates for Enzyme Biosensors

Carbon is a commonly used solid electrode material as a substrate for biosensors, particularly in the form of glassy carbon, due to its wide positive potential window, mechanical stability, and low porosity. Carbon film, carbon composite and graphite are other substrates which have been investigated. Carbon film electrodes made from carbon film electrical resistors have been extensively employed by us as electrode substrate [43], obtained by pyrolytically coating a ceramic cylindrical substrate with a thin carbon layer.

Characterisation has been carried out by voltammetry, electrochemical impedance spectroscopy (EIS), and most recently by atomic force microscopy (AFM) and confocal Raman spectroscopy [44]. A typical AFM image of the carbon film surface is shown in Fig. 6.5a, demonstrating the roughness arising from the ceramic substrate and Fig. 6.5b, clearly shows the graphitic “D” and “G” bands which appear for thicker carbon films, corresponding to lower electrical resistance, the graphitic nature being essential for electrochemical behaviour similar to that of bulk conducting carbon electrodes.

The understanding of the electrode processes occurring is reliant on electrochemical data or on microscopic/spectroscopic information. In order to obtain further complementary, simultaneous information, investigations have begun with platinum or gold piezoelectric quartz crystals further coated with carbon [45].

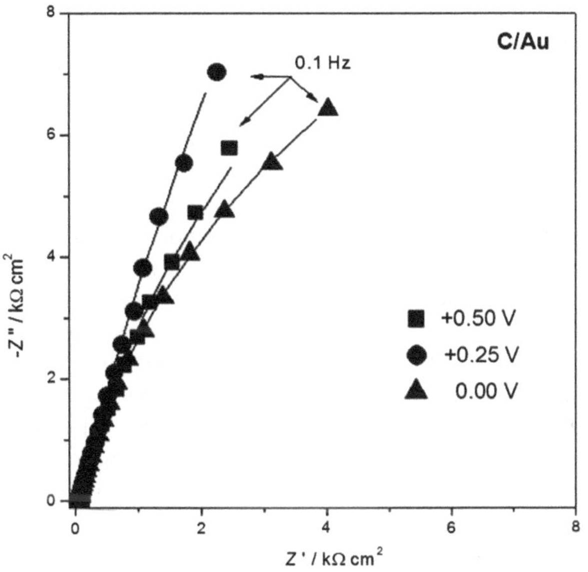

Fig. 6.6 Complex plane electrochemical impedance spectra of a carbon-film coated piezoelectric quartz crystal; potentials are versus saturated calomel electrode. [From Ref. 45 with permission]

These can then be mounted into an electrochemical quartz crystal microbalance (EQCM) so that mass and viscoelastic changes can be monitored as a function of the applied potential or current; the sensitivity is of the order of 4 ng cm^{-2}. Besides AFM, EIS was also employed for characterisation, as illustrated in Fig. 6.6. which shows the response at the carbon film coated quartz crystal at various potentials and is similar to that at other forms of carbon. The films are robust and are promising for use to study a variety of adsorption and deposition processes associated with the EQCM, in particular for biosensor applications.

6.3.2 Redox Mediator-Modified Electrodes

Redox mediators based on films of metal hexacyanoferrates [46, 47], and on polyphenazine polymers [48] are amongst the most common. The former are prepared in the appropriate solution as modifier films by electrodeposition or autocatalytic deposition on electrode substrates. The latter are formed by electropolymerisation using potential cycling, and retain the redox properties of the corresponding monomers—the fact that the redox centres are fixed on the electrode surface aids in increasing the effective concentration (and thus sensitivity) as well as permitting an enzyme layer to be constructed on top of the film. A review of the use of polyphenazines in sensors and biosensors has recently been published [48]. Poly(neutral red) (PNR) was used in the construction of a number of biosensors using oxidase enzymes, described together with a discussion of the mechanism of action in [49, 50], and application in a mixture with carbon nanotubes to enhance the sensitivity [51]. The main redox polymer peak is around −0.4 V versus saturated calomel electrode (SCE) and this dictates its use as a successful redox mediator; such films are also robust and stable with time. Brilliant cresyl blue has also been recently described as monomer molecules trapped within a sol–gel film (see below) for a peroxidase biosensor [52] and as a polymer film for redox-mediated biosensor applications [53].

One of the main questions to explore is how polymer nucleation and growth are influenced by the electrode substrate and by the electrolyte solution. In experiments linked to some studies undertaken in room temperature ionic liquids, the formation of PNR was also investigated. Room temperature ionic liquids (RTILs) are excellent for electrochemistry in the sense that they are solvent and electrolyte at the same time; however, besides being hygroscopic, their high viscosity makes diffusion much slower and thus the electropolymerisation rate, polymer morphology and structure, were significantly influenced as shown voltammetrically [54]. This illustrates the potential of this approach for tuning the nanostructure of the polymer films by changing the polymerisation medium, and opens up interesting opportunities for the future.

6.3.3 Enzyme Immobilisation Strategies for Biosensors

The immobilisation of enzymes in the enzyme layer should be carried out to allow:

- easy access of the enzyme substrate to the enzyme active centre whilst allowing the active site to change its size and conformation during interaction,
- easy release of the reaction products and their diffusion to a location where they are detected by oxidation or reduction.

Thus, immobilisation should not lead to enzyme deactivation, make access to the enzyme active site difficult or let enzyme leach into the surrounding solution. The more common methodology for immobilisation of enzymes on electrode surfaces involves cross-linking with glutaraldehyde, aided by the protein bovine serum albumin (BSA) in order to increase the interaction with the enzyme molecules. In this section two other strategies will be described.

Enzyme encapsulation in sol–gel oxysilane networks, formed from oxysilane precursors which undergo condensation reactions to form a polymer network, produces nanocages within which the enzyme is encapsulated. Rather than traditional precursors such as methyltrimethoxysilane (MTMOS) the use of 3-aminopropyltriethoxysilane (APTOS) has been explored. This monomer contains an amino group that is able to interact with the enzyme to reduce any leaching. The monomer 3-glycidoxypropyltrimethoxysilane (GOPMOS) has also been investigated and which contains a long spacer arm with an epoxy group, thus increasing the size of the nanocages so that the conformation of the enzyme molecule can more easily change as needed during reaction [55].

Fig. 6.7 AFM topographical images in air of sol–gel deposited onto HOPG: **a** MTMOS, **b** APTOS, and **c** GOPMOS. [From Ref. 56 with permission]

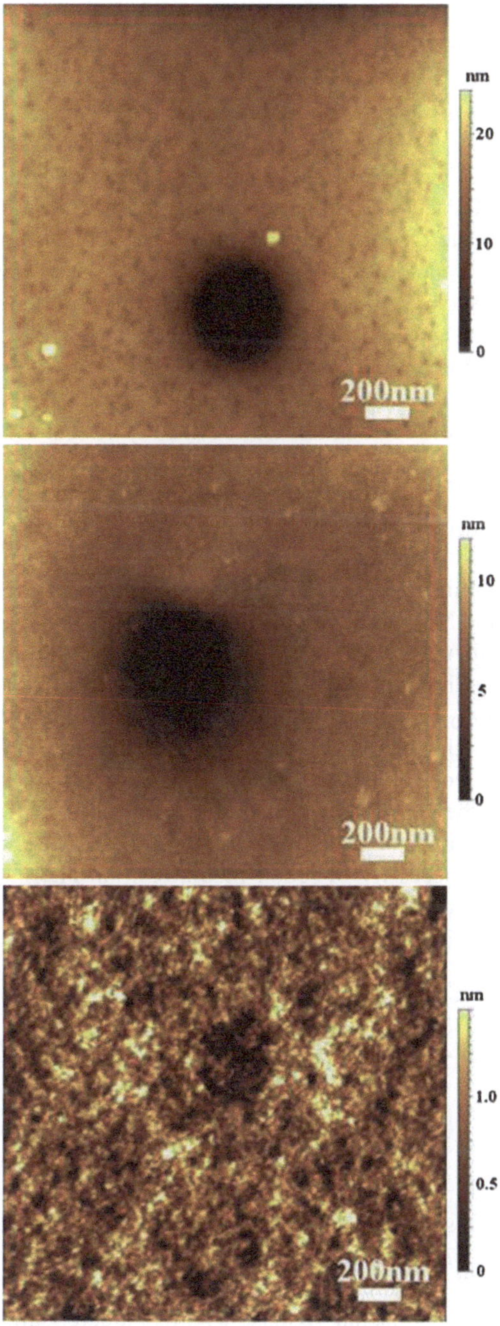

It was shown that significant benefits can be obtained from the use of APTOS/GOPMOS mixtures, and which avoid the cracking that can occur with sol–gel networks on drying.

Investigations of the morphology were carried out by AFM without mediator film [56], see Fig. 6.7, and with poly (neutral red) [57]. It was convincingly shown that the pores were smaller and less numerous in both cases using the APTOS/GOPMOS combination. Thus, in order to prevent leaching, thin external coatings of inert, semipermeable polymers were applied, characterised by voltammetry and EIS and tested [58]. The best of these was polyurethane- although biosensor response was one half of what it was without the polymer layer, it remained almost constant after 2 months of use.

Other recent examples of the formation of oxysilane sol–gel biosensors and characterisation of the assembly build-up using electrochemical and AFM techniques are found in the recent literature for glucose [59], lactate [60] and uric acid [61].

A second enzyme immobilisation strategy uses chitosan matrices and multi-walled carbon nanotubes (MWCNT) functionalised with carboxylate groups [62]. Chitosan is a natural linear polysaccharide from glucoseamine, synthesized from chitin, which is found in crustaceans. For the best biosensor, mixtures of chitosan and MWCNT were placed onto the surface of a carbon-epoxy composite electrode; this was followed by a mixture of enzyme with BSA and as cross-linking agent, a combination of 1-ethyl-3-(3-dimethylaminopropyl) carbodiimide followed by N-hydroxysuccinimide. Covalent binding of the chitosan as well as the MWCNT on the electrode surface was achieved. This is a robust structure which, with glucose oxidase as model enzyme, gives a very good response to glucose [63]. The presence of the multiwalled nanotubes increases the sensitivity and enables the sensor to be used at a potential of –0.2 V versus SCE (peroxide reduction) without the necessity of redox mediators. An alternative strategy involves modifying the carbon nanotubes in the chitosan matrix directly with redox mediator, in [64] with a pentacyaneferrate derivative. .

6.3.4 Direct Electron Transfer and Layer-by-Layer Self-Assembly

In a future, ideal enzyme biosensor, the enzymes would be immobilised in such a way that all the enzyme molecules react with enzyme substrate diffusing easily from solution into the enzyme layer and the products of the enzyme reaction reach the sensor transducer surface without any diffusion limitations, which leads to efficiencies approaching 100 %. Ideally, there should be no redox mediator and direct electron transfer occurs. Two recent examples are a biosensor platform based on lactate oxidase where characterisation was done by AFM and SECM [65] and glucose oxidase using functionalised nanotubes within a dihexadecylphosphate film (DHP) [66]—in this latter case, SEM shows clearly CNTs distributed homogeneously in the DCP film, Fig. 6.8.

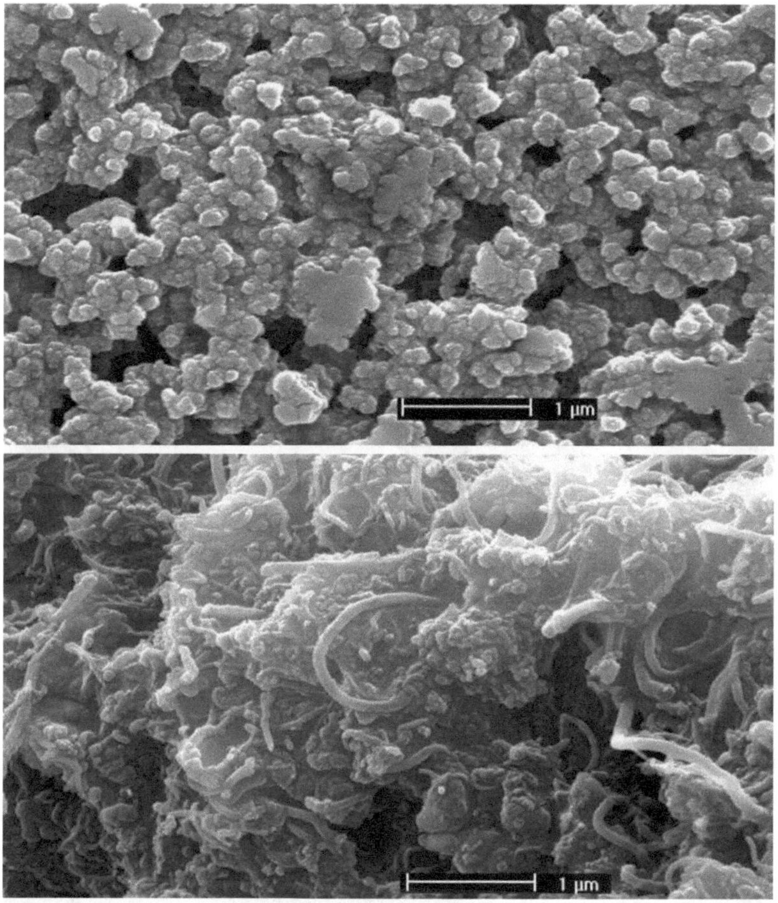

Fig. 6.8 SEM images of DHP and CNTs-DHP on the surface of GC electrode.SEM of CNT-DHP. [From Ref. 66 with permission]

Glucose oxidase was also adsorbed onto hydrophilic and positively-charged plasma-polymerised thin film surfaces, being characterised by AFM, QCM and electrochemistry [67], also functioning as a glucose biosensor.

Molecularly-thick sensor layers reduce diffusion limitations, and the construction of multilayer sensors self-assembled using the layer-by-layer technique is one of the ways in which this can be achieved.

Two examples of this layer-by-layer strategy are illustrated. In the first, electrochemical enzyme biosensors for glucose were prepared. Indium tin oxide (ITO) glass electrodes were modified by self-assembly using an immersion technique with up to three bilayers of polyamidoamine (PAMAM) dendrimers containing gold nanoparticles of diameter ~3 nm and poly(vinyl sulfonate) (PVS). The gold

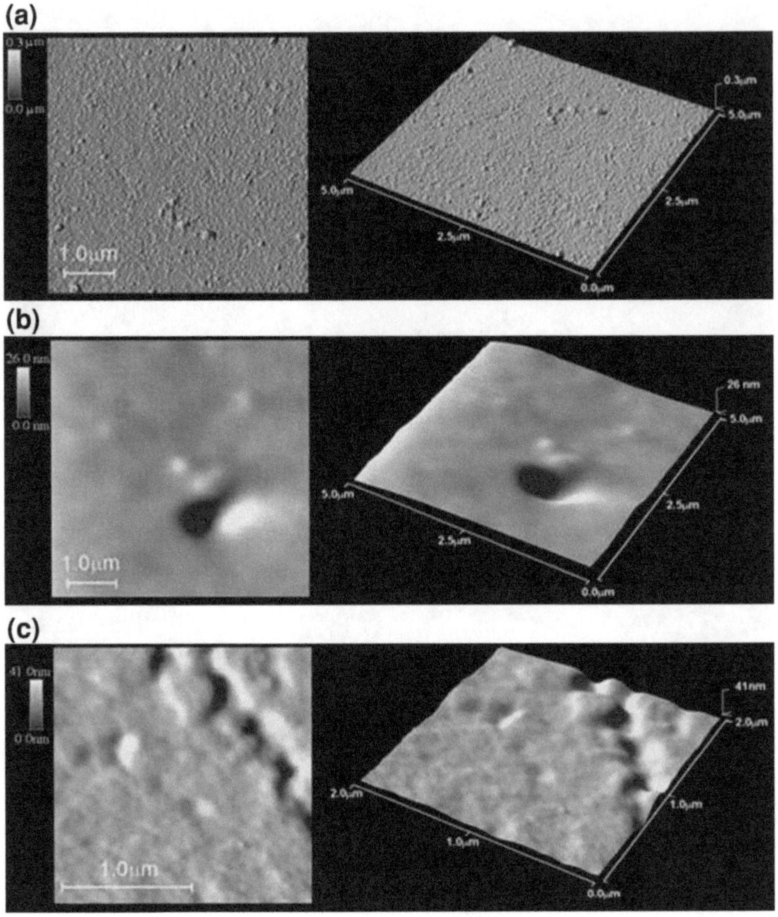

Fig. 6.9 AFM topographical images in 2D and 3D for AuQCM-(MPS(−)/PPD(+)/$\{HA/Mb\}_n$, with n = **a** 2 **b** 4 and **c** 6 bilayers. [From Ref. 70 with permission]

nanoparticles were covered with cobalt hexacyanoferrate that functioned as a redox mediator, allowing the modified electrode to be used to detect hydrogen peroxide, at 0.0 V versus SCE [68]. Glucose oxidase enzyme was then immobilised by crosslinking with glutaraldehyde and BSA. The optimised biosensor, with three bilayers, has high sensitivity and operational stability, with a detection limit of 6.1 μM and showed good selectivity against interferents.

Secondly, other research has involved the probing of immobilised biological species. The layer-by-layer self-assembled film construction of the biocompatible polymer hyaluronic acid (HA) and single heme redox protein, myoglobin (Mb) was done on gold electrode substrates, both gold quartz crystal electrodes (EQCM) and bulk gold (Au(bulk)) electrodes [69]. The electrochemical properties of the hyaluronic acid/myoglobin films ($\{HA/Mb\}_n$) were investigated after each deposition step using cyclic voltammetry, quartz crystal microbalance and electrochemical impedance spectroscopy [70, 71]. AFM was particularly helpful in demonstrating the number of bilayers necessary to enable the substrate surface to be completely covered, see Fig. 6.9. Other investigations have shown this to be generally the case. The voltammetric response demonstrated the presence of free haemoglobin in the adsorbed multilayer films.

The layer-by-layer technique with the negative "hyaluronate" form of the polymer HA, can be applied for the immobilisation of other positively charged biomolecules, as shown in [72]. This enables the inexpensive and easy construction of very stable films, with the use of very small amounts of enzyme or protein which augurs well for future sensor construction.

6.4 Conclusion

The use of complementary surface analysis tools to understand the electrochemical phenomena for DNA and enzyme electrochemical biosensors is of paramount importance for the design of improved, more efficient and robust sensors.

The electrochemical DNA-biosensor is a complementary tool for the study of biomolecular interaction mechanisms of compounds with DNA, enabling the screening and evaluation of the effect caused to DNA by health hazardous compounds and oxidising substances. The characterization of a DNA-electrochemical biosensor provides very relevant information because the mechanisms of DNA-hazard compound interaction at charged interfaces mimic better the in vivo situation, opening wide perspectives using a particularly sensitive and selective method for the detection of specific interactions.

The goal of achieving direct electron transfer for enzyme biosensors requires detailed nanostructuring of the surface and orienting of the enzyme molecules to maximise efficiency. Microscopic techniques play a very significant role in enabling this to be carried out. Using these tools, particular importance can be expected to be given to layer-by-layer assembly in the future.

Acknowledgments Financial support from Fundação para a Ciência e Tecnologia (FCT), and CEMUC-R (Research Unit 285), is gratefully acknowledged.

References

1. Brett, C.M.A.: Novel sensor devices and monitoring strategies for green and sustainable chemistry processes. Pure Appl. Chem. **79**, 1969–1980 (2007)
2. Paleček, E., Fojta, M., Jelen, F., Vetterl, V.: Electrochemical analysis of nucleic acids. In: Bard, A.J., Stratmann, M. (eds.) The Encyclopedia of Electrochemistry, vol. 9, Ch 12: 365–429, Wiley-VCH, Germany (2002) and references therein.

3. Fojta, M.: Detecting DNA damage with electrodes. In: Palecek, E., Sheller, F., Wang, J., (eds) Perspectives in Bioanalysis, vol. 1: 385–431, Elsevier, Amsterdam (2005)
4. Oliveira-Brett, A.M., Diculescu, V.C., Chiorcea-Paquim, A.M., Serrano, S.H.P.: DNA-electrochemical biosensors for investigating DNA damage. In: Alegret S, Merkoçi A (eds.) Comprehensive Analytical Chemistry, vol. 49, Ch 4: 413–437, Elsevier, Amsterdam (2007) and references therein
5. Oliveira-Brett, A.M.: Electrochemistry for probing DNA damage. In: Grimes, C.A., Dickey, E.C., Pishko, M.V. (eds.) Encyclopedia of Sensors, vol. 3: pp. 301–314, American Scientific Publishers, Los Angeles (2006) and references therein.
6. Labuda, J., Oliveira-Brett, A.M., Evtugyn, G., Fojta, M., Mascini, M., Ozsoz, M., Palchetti, I., Paleček, E., Wang, J.: Electrochemical nucleic acid-based biosensors: concepts, terms, and methodology. (IUPAC Technical Reports and Recommendations). Pure Appl. Chem. **82**, 1161–1187 (2010)
7. Oliveira-Brett, A,M., Diculescu, V.C., Chiorcea-Paquim, A.M., Serrano, S.H.P.: DNA-electrochemical biosensors for investigating DNA damage. In Alegret S, Merkoçi A (eds) Comprehensive Analytical Chemistry, vol. 49, Proc. 28, e203-e205, Proc. 29, e207-e211, Elsevier, Amsterdam, (2007)
8. Ravera, M., Bagni, G., Mascini, M., Osella, D.: DNA-metallodrugs interactions signaled by electrochemical biosensors: An overview. Bioinorg. Chem. Appl, Article ID 91078, 1–11 (2007)
9. Oliveira, S.C.B., Oliveira-Brett, A.M.: Boron doped diamond electrode pre-treatments effect on the electrochemical oxidation of dsDNA, DNA bases, nucleotides, homopolynucleotides and biomarker 8-oxoguanine. Electrochim. Acta **648**, 60–66 (2010)
10. Halliwell, B., Gutteridge, J.M.C.: Biologically relevant metal ion-dependent hydroxyl radical generation—an update. FEBS Lett. **307**, 108–112 (1992)
11. Halliwell, B., Gutteridge, J.M.C.: Free radicals in biology and medicine. Oxford University, UK (1999)
12. Klungland, A., Bjelland, S.: Oxidative damage to purines in DNA: Role of mammalian Ogg1. DNA Repair **6**, 481–488 (2007)
13. Shibutani, S., Takeshita, M., Grollman, A.P.: Insertion of specific bases during DNA-synthesis past the oxidation-damaged base 8-oxodg. Nature **349**, 431–434 (1991)
14. Bjelland, S., Seeberg, E.: Mutagenicity, toxicity and repair of DNA base damage induced by oxidation. Mutat. Res. Fund. Mol. Mech. Mutat. **531**, 37–80 (2003)
15. Kasai, H.: Analysis of a form of oxidative DNA damage, 8-hydroxy-2′-deoxyguanosine, as a marker of cellular oxidative stress during carcinogenesis. Mutat. Res. Rev. Mutat. **387**, 147–163 (1997)
16. Halliwell, B.: Oxygen and nitrogen are pro-carcinogens. Damage to DNA by reactive oxygen, chlorine and nitrogen species: measurement, mechanism and the effects of nutrition. Mutat. Res. Genet. Toxicol. Environ. **443**, 37–52 (1999)
17. Borish, E.T., Cosgrove, J.P., Church, D.F., Deutsch, W.A., Pryor, W.A.: Cigarette tar causes single-strand breaks in DNA. Biochem. Biophys. Res. Commun. **133**, 780–786 (1985)
18. Olinski, R., Gackowski, D., Foksinski, M., Rozalski, R., Roszkowski, K., Jaruga, P.: Oxidative DNA damage: assessment of the role in carcinogenesis, atherosclerosis, and acquired immunodeficiency syndrome. Free Radic. Bio. Med. **33**, 192–200 (2002)
19. Oliveira-Brett, A.M., Piedade, J.A.P., Serrano, S.H.P.: Electrochemical oxidation of 8-oxoguanine. Electroanal. **12**, 969–973 (2000)
20. Diculescu, V.C., Piedade, J.A.P., Oliveira-Brett, A.M.: Electrochemical behaviour of 2,8-dihydroxyadenine at a glassy carbon electrode. Bioelectrochemistry **70**, 141–146 (2007)
21. Oliveira-Brett, A.M., Vivan, M., Fernandes, I.R., Piedade, J.A.P.: Electrochemical detection of in situ adriamycin oxidative damage to DNA. Talanta **56**, 959–970 (2002)
22. Piedade, J.A.P., Fernandes, I.R., Oliveira-Brett, A.M.: Electrochemical sensing of DNA-adriamycin interactions. Bioelectrochemistry **56**, 81–83 (2002)

23. Diculescu, V.C., Vivan, M., Oliveira-Brett, A.M.: Voltammetric behavior of antileukemia drug glivec. Part III: In situ DNA oxidative damage by the glivec electrochemical metabolite. Electroanalysis **18**, 1963–1970 (2006)
24. Oliveira-Brett, A.M., Macedo, T.R.A., Raimundo, D., Marques, M.H., Serrano, S.H.P.: Voltammetric behaviour of mitoxantrone at a DNA-biosensor. Biosens. Bioelectron. **13**, 861–867 (1998)
25. Oliveira-Brett, A.M., Serrano, S.H.P., Macedo, T.A., Raimundo, D., Marques, M.H., LaScalea, M.A.: Electrochemical determination of carboplatin in serum using a DNA-modified glassy carbon electrode. Electroanalysis **8**, 992–995 (1996)
26. La-Scalea, M.A., Serrano, S.H.P., Ferreira, E.I., Oliveira-Brett, A.M.: Voltammetric behavior of benznidazole at a DNA-electrochemical biosensor. J. Pharm. Biomed. **29**, 561–568 (2002)
27. Oliveira-Brett, A.M., Serrano. S.H.P., La-Scalea, M.A., Gutz, I.G.R., Cruz, M.L.: Studies on the mechanism of interaction of in situ produced nitroimidazole reduction derivatives with DNA using an electrochemical DNA-biosensor. In: Packer, L. (ed.) Methods in Enzymology, 300 Part B: 314–321, Academic Press, Cleveland (1999)
28. Oliveira, S.C.B., Vivan, M., Oliveira-Brett, A.M.: Electrochemical behavior of thalidomide at a glassy carbon electrode. Electroanalysis **20**, 2429–2434 (2008)
29. Oliveira, S.C.B., Chiorcea-Paquim, A.M., Ribeiro, S.M., Melo, A.T.P., Vivan, M., Oliveira-Brett, A.M.: In situ electrochemical and AFM study of thalidomide-DNA interaction. Bioelectrochemistry **76**, 201–207 (2009)
30. Diculescu, V.C., Chiorcea-Paquim, A.-M., Tugulea, L., Vivan, M., Oliveira-Brett, A.M.: Interaction of imatinib with liposomes: voltammetric and AFM characterization. Bioelectrochemistry **74**, 278–288 (2009)
31. Corduneanu, O., Chiorcea-Paquim, A.-M., Garnett, M., Oliveira-Brett, A.M.: Lipoic acid—palladium complex interaction with DNA, voltammetric and AFM characterization. Talanta **77**, 1843–1853 (2009)
32. Chiorcea-Paquim, A.-M., Corduneanu, O., Oliveira, S.C.B., Diculescu, V.C., Oliveira-Brett, A.M.: Electrochemical and AFM evaluation of hazard compounds-DNA interaction. Electrochim. Acta **54**, 1978–1985 (2009)
33. Oliveira, S.C.B., Oliveira-Brett, A.M.: DNA-electrochemical biosensors: AFM surface characterisation and application to detection of in situ oxidative damage to DNA. Comb. Chem. High Throughput Screen. (CC&HTS) **13**, 628–640 (2010)
34. Oliveira, S.C.B., Oliveira–Brett, A.M.: In situ evaluation of chromium-DNA damage using a DNA-electrochemical. Biosensor. Anal. Bioanal. Chem. **398**, 1633–1641 (2010)
35. Corduneanu, O., Chiorcea-Paquim, A.-M., Fiuza, S.M., Marques, M.P.M., Oliveira-Brett, A.M.: Polynuclear palladium complexes with biogenic polyamines: AFM and voltammetric characterization. Bioelectrochemistry **78**, 97–105 (2010)
36. Corduneanu, O., Chiorcea-Paquim, A.-M., Diculescu, V., Fiuza, S.M., Marques, M., Oliveira-Brett, A.M.: Interaction of DNA with palladium chelates of biogenic polyamines - AFM and voltammetric characterization. Anal. Chem. **82**, 1245–1252 (2010)
37. Pontinha, A.D.R., Jorgem S.M.A., Chiorcea-Paquim, A.-M., Diculescu, V.C., Oliveira Brett, A.M.: In situ evaluation of anticancer drug methotrexate-DNA interaction using a DNA-electrochemical biosensor and AFM characterization. Phys. Chem. Chem. Phys. (PCCP) **13** (12): 5227–5234 (2011)
38. Satana, H.E., Oliveira-Brett, A.M.: In situ evaluation of fludarabine-DNA interaction using a DNA-electrochemical biosensor. Int. J. Electrochem. vol. 2011, Article ID 340239, 8 pages, (2011) doi:10.4061/2011/340239.
39. Satana, H.E., Pontinha, A.D.R., Diculescu, V.C., Oliveira-Brett, A.M.: Nucleoside analogue electrochemical behaviour and in situ evaluation of DNA-clofarabine interaction. Bioelectrochemistry (2011) doi:10.1016/j.bioelechem.2011.07.004
40. Diculescu, V.C., Chiorcea-Paquim, A.-M., Eritja, R., Oliveira-Brett, A.M.: Thrombine binding aptamer quadruplex formation: AFM and voltammetric characterization. J. Nucleic Acids **1**, 1–8 (2010)

41. Diculescu, V.C., Chiorcea-Paquim, A.-M., Eritja, R., Oliveira-Brett, A.M.: Evaluation of the structure-activity relationship of thrombin with thrombin binding aptamers by voltammetry and atomic force microscopy. J. Electroanal. Chem. **656**, 159–166 (2011)
42. Casero, E., Vazquez, L., Parra-Alfambra, A.M., Lorenzo, E.: AFM; SECM and QCM as useful analytical tools in the characterization of enzyme-based bioanalytical platforms. Analyst **135**, 1878–1903 (2010)
43. Brett, C.M.A., Angnes, L., Liess, H.-D.: Carbon film resistors as electrodes: voltammetric properties and application in electroanalysis. Electroanalysis **13**, 765–769 (2001)
44. Gouveia-Caridade, C., Soares, D.M., Liess, H.-D., Brett, C.M.A.: Electrochemical, morphological and microstructural characterization of carbon film resistor electrodes for application in electrochemical sensors. Appl. Surf. Sci. **254**, 6380–6389 (2008)
45. Pinto, E.M., Gouveia-Caridade, C., Soares, D.M., Brett, C.M.A.: Electrochemical and surface characterization of carbon-film-coated piezoelectric quartz crystals. Appl. Surf. Sci. **255**, 8084–8090 (2009)
46. Ricci, F., Palleschi, G.: Sensor and biosensor preparation, optimisation and applications of Prussian Blue modified electrodes. Biosens. Bioelectron. **21**, 389–407 (2005)
47. Pauliukaite, R., Florescu, M., Brett, C.M.A.: Characterization of cobalt and copper hexacyanoferrate modified carbon film electrodes for redox mediated biosensors. J. Solid State Electrochem. **9**, 354–362 (2005)
48. Pauliukaite, R., Ghica, M.E., Barsan, M.M., Brett, C.M.A.: Phenazines and polyphenazines in electrochemical sensors and biosensors. Anal. Lett. **43**, 1588–1608 (2010)
49. Pauliukaite, R., Ghica, M.E., Barsan, M., Brett, C.M.A.: Characterisation of poly(neutral red) modified carbon film electrodes; application as a redox mediator for biosensors. J. Solid State Electrochem. **11**, 899–908 (2007)
50. Pauliukaite, R., Brett, C.M.A.: Poly(neutral red): electrosynthesis, characterisation and application as a redox mediator. Electroanalysis **20**, 1275–1285 (2008)
51. Carvalho, R.C., Gouveia-Caridade, C., Brett, C.M.A.: Glassy carbon electrodes modified by multiwalled carbon nanotubes and poly(neutral red). A comparative study of different brands and application to electrocatalytic ascorbate determination. Anal. Bioanal. Chem. **398**, 1675–1685 (2010)
52. Peng, Y.Y., Upadhyay, A.K., Chen, S.M.: Analytical biosensing of hydrogen peroxide on brilliant cresyl blue/multiwalled carbon nanotubes modified glassy carbon electrode. Electroanalysis **22**, 463–470 (2010)
53. Ghica, M.E., Brett, C.M.A.: Poly(brilliant cresyl blue) modified glassy carbon electrodes: electrosynthesis, characterisation and application in biosensors. J. Electroanal. Chem. **629**, 35–42 (2009)
54. Pauliukaite, R., Doherty, A.P., Murnaghan, K.D., Brett, C.M.A.: Application of some room temperature ionic liquids in the development of biosensors at carbon film electrodes. Electroanalysis **20**, 485–490 (2008)
55. Pauliukaite, R., Brett, C.M.A.: Characterization of novel glucose oxysilane sol-gel electrochemical biosensors with copper hexacyanoferrate mediator. Electrochim. Acta **50**, 4973–4980 (2005)
56. Pauliukaite, R., Chiorcea-Paquim, A.-M., Oliveira-Brett, A.M., Brett, C.M.A.: Electrochemical, EIS and AFM characterisation of biosensors: trioxysilane sol-gel encapsulated glucose oxidase with two different redox mediators. Electrochim Acta **52**, 1–8 (2006)
57. Chiorcea-Paquim, A.-M., Pauliukaite, R., Brett, C.M.A., Oliveira-Brett, A.M.: AFM nanometer surface morphological study of in situ electropolymerized neutral red redox mediator oxysilane sol-gel encapsulated glucose oxidase electrochemical biosensors. Biosens. Bioelectron. **24**, 297–305 (2008)
58. Pauliukaite, R., Schoenleber, M., Vadgama, P., Brett, C.M.A.: Development of electrochemical biosensors based on sol-gel enzyme encapsulation and protective polymer membranes. Anal. Bioanal. Chem. **390**, 1121–1131 (2008)

59. Barbadillo, M., Casero, E., Petit-Dominguez, M.D., Pariente, F., Lorenzo, E., Vazquez, L.: Surface study of the building steps of enzymatic sol-gel biosensors at the micro- and nano-scales. J. Sol-Gel. Sci. Technol. **58**, 452–462 (2011)
60. Parra-Alfambra, A.M., Casero, E., Petit-Domingez, M.D., Barbadillo, M., Pariente, F., Vazquez, L., Lorenzo, E.: New nanostructured electrochemical biosensors based on three-dimensional (3-mercaptopropyl)-trimethoxysilane network. Analyst **136**, 340–347 (2011)
61. Ahuja, T., Tanwar, V.K., Mishra, S.K., Kumar, D., Biradar, A.M., Rajesh,: Immobilization of uricase enzyme on self-assembled gold nanoparticles for application in uric acid biosensor. J. Nanosci. Nanotechnol. **11**: 4692–4701 (2011)
62. Pauliukaite, R., Ghica, M.E., Fatibello-Filho, O., Brett, C.M.A.: A comparative study of different crosslinking agents for the immobilization of functionalized carbon nanotubes within a chitosan film supported on a graphite-epoxy composite electrode. Anal. Chem. **81**, 5364–5372 (2009)
63. Ghica, M.E., Pauliukaite, R., Fatibello-Filho, O., Brett, C.M.A.: Application of functionalised carbon nanotubes immobilised into chitosan films in amperometric enzyme biosensors. Sens. Actuat. B **142**, 308–315 (2009)
64. Parra-Alfambra, A.M., Casero, E., Ruiz, M.A., Vazquez, L., Pariente, F., Lorenzo, E.: Carbon nanotubes/pentacyaneferrate-modified chitosan nanocomposites platforms for reagentless glucose biosensing. Anal. Bioanal. Chem. **401**, 883–889 (2011)
65. Parra, A., Casero, E., Vazquez, L., Jin, J., Pariente, F., Lorenzo, E.: Microscopic and voltammetric characterization of bioanalytical platforms based on lactate oxidase. Langmuir **22**, 5443–5450 (2006)
66. Janegitz, B.C., Pauliukaite, R., Ghica, M.E., Brett, C.M.A., Fatibello-Filho, O.: Direct electron transfer of glucose oxidase at glassy carbon electrode modified with functionalized carbon nanotubes within a dihexadecylphosphate film. Sens. Actuat. B **158**, 411–417 (2011)
67. Mugurama, H., Kase, Y., Murata, N., Matsumura, K.: Adsorption of glucose oxidase onto plasma-polymerized film characterized by atomic force microscopy, quartz crystal microbalance, and electrochemical measurement. J. Phys. Chem. B **110**, 26033–26039 (2006)
68. Crespilho, F.N., Ghica, M.E., Gouveia-Caridade, C., Oliveira Jr. O.N., Brett, C.M.A.: Enzyme immobilisation on electroactive nanostructured membranes (ENM): optimised architectures for biosensing. Talanta 76:922–928 (2008)
69. Barsan, M.M., Pinto, E.M., Brett, C.M.A.: Interaction between myoglobin and hyaluronic acid in layer-by-layer structures - an electrochemical study. Electrochim. Acta **55**, 6358–6366 (2010)
70. Pinto, E.M., Barsan, M.M., Brett, C.M.A.: Mechanism of formation and construction of self-assembled myoglobin/hyaluronic acid multilayer films—an electrochemical QCM, impedance and AFM study. J. Phys. Chem. B **114**, 15354–15361 (2010)
71. Barsan, M.M., Pinto, E.M., Brett, C.M.A.: Methylene blue and neutral red electropolymerisation on AuQCM and on modified AuQCM electrodes: An electrochemical and gravimetric study. Phys. Chem. Chem. Phys. **13**, 5462–5471 (2011)
72. Gutierrez-Sanchez, C., Olea, D., Marques, M., Fernandez, V.M., Pereira, I.A.C., Velez, M., De Lacey, A.L.: Oriented immobilization of a membrane-bound hydrogenase onto an electrode for direct electron transfer. Langmuir **27**, 6449–6457 (2011)

Chapter 7
Electrochemical-Surface Plasmon Resonance: Concept and Bioanalytical Applications

Danielle C. Melo Ferreira, Renata Kelly Mendes and Lauro Tatsuo Kubota

Abstract The combination of surface plasmon resonance (SPR) and electrochemical methods has become a powerful technique for simultaneous observation of optical and electrochemical properties at substrate/electrolyte interfaces. The fundamental aspects of the electric potential effects on surface plasmons are introduced and the use and applications of this combined electrochemical and optical technique are discussed. Electrochemical-Surface Plasmon Resonance (ESPR) has several advantages, such as: spatial resolution, which is particularly attractive for studying heterogeneous reactions; optical properties of reactive species that may assist identification action mechanisms, and high surface sensitivity for studying surface binding of the reaction species. The electrochemistry-SPR spectroscopy technique has also been used for many applications, including bio-analytical systems that will be further described in more detail.

D. C. Melo Ferreira
Laboratório de Microfabricação—Laboratório Nacional de Nanotecnologia, CNPEM, Caixa Postal 6192, Campinas, São Paulo 13083-970, Brazil

R. K. Mendes
Faculdade de Química, Pontifícia Universidade Católica de Campinas, Campinas, São Paulo, Brazil

L. T. Kubota (✉)
Instituto de Química, Universidade Estadual de Campinas—UNICAMP, P.O. Box 6154, Campinas, São Paulo 13083-970, Brazil
e-mail: kubota@iqm.unicamp.br

F. N. Crespilho (ed.), *Nanobioelectrochemistry*, DOI: 10.1007/978-3-642-29250-7_7,

7.1 Introduction

Since surface plasmon resonance was first proposed in the 1980s to be used as a label-free technique for direct monitoring of specific antibody and antigen interactions, the popularity of this technology for biosensor applications has grown rapidly and resulted in a vast array of platforms available for research laboratory settings [1]. SPR can in situ detect the concentrations of biomolecules during the binding process, being a powerful surface-sensitive characterization method. Furthermore, the kinetic data including the equilibrium constant, the association and dissociation parameters between biomolecules can also be obtained by simulating SPR kinetic curves [2]. Today, SPR is one of the most frequently used techniques to monitor interfacial reactions at the solid/liquid interface. The methodology is suitable for several types of analysis. However, its sensitivity chiefly depends on the mass change on the SPR chip [3].

The combination of the electrochemical and SPR techniques can provide multidimensional information on the properties and characteristics of the electrode surface and has proven to be useful. Hence, electrochemical methods, such as cyclic voltammetry (CV) and electrochemical impedance spectroscopy (EIS), which present advantages such as: high sensitivity and simplicity, are very effective to monitor the characteristics of electrode/electrolyte interfaces [4].

Electrochemical-SPR measurements have been used to characterize structural and optical properties involving the analysis of biosensors. The simultaneous approach is logical since both methods are—from an instrumental point of view—highly complementary and have found widespread applications in different research domains, including studies of the electrochemical double layer, the investigation of the electrochemical doping process, as well as electrical field enhanced studies [5, 6].

Some studies express that SPR is an efficient to examine the optical property changes during the electropolymerization of electroactive monomers. There are many advantages by combining electrochemistry and SPR measurements to evaluate polymer formation. For example, the refractive index or film thickness changes of polymers during electropolymerization and ion incorporation processes upon doping and dedoping of the films can be elucidated by the combination of both techniques to develop a platform to immobilize biomolecules [7].

Based on the wide field of the SPR applications associated with investigating these processes, as well as on the various promising future uses and applications, this work presents and discusses the basic principles of the use of the SPR technique in the investigation of electrochemical processes, featuring the relevance of the concomitant analysis of optical and electrochemical processes described as cases studies.

7.2 Surface Plasmon Resonance

The physical phenomenon of surface plasmon resonance observed by Wood in the beginning of the twentieth century has found its way into practical applications in sensitive detectors, capable of detecting sub-monomolecular coverage. Wood observed a pattern of different dark and light bands in the reflexcited light, when he shone polarized light on a mirror with a diffraction grating on its surface. Physical interpretation of the phenomenon was initiated by Lord Rayleigh, and further refined by Fano, but a complete explanation of the phenomenon was not possible until 1968, when Otto, and in the same year Kretschmann and Raether reported the excitation of surface plasmons. Since then, applications of SPR-based sensors to biomolecular interaction monitoring are developed [8–13].

Surface plasmon resonance is an optical technical that measure the resonance coupling of incident light to the propagating surface plasmon. A planar structure consisting of a thick metal film sandwiched between two semi-infinite dielectrics supports two independent surface plasmons at the opposite boundaries of the metal film. If the metal film is thin, coupling between the surface plasmons at opposite boundaries of the metal film can occur, giving rise to mixed modes of electromagnetic field symmetric and antisymmetric surface plasmons. The symmetric surface plasmon exhibits a propagation constant and attenuation, which both increase with increasing metal film thickness. The propagation constant and attenuation of the antisymmetric surface plasmon decrease with increasing thickness of the metal film. The symmetric surface plasmon exhibits a lower attenuation than its antisymmetric counterpart, and therefore it is referred to as a long-range surface plasmon, whereas the antisymmetric mode is referred to as a short-range surface Plasmon (Fig. 7.1) [10, 14–17].

Surface plasmon resonance is an excellent method to monitor changes of the refractive index in the near vicinity of the metal surface. When the refractive index changes, the angle at which the intensity minimum is observed will shift as indicated in Figs. 7.2a and b, where (a) depicts the original plot of reflected light intensity versus incident angle and (b) indicates the plot after the change in refractive index. Surface plasmon resonance is not only suited to measure the difference between these two states, but can also monitor the change in time, if one follows in time the shift of the resonance angle at which the dip is observed. The Fig. 7.2a depicts the shift of the dip in time, a so-called sensorgram. If this change is due to a molecular interaction, the kinetics of the interaction can be studied in real time.

7.3 Kinetic Parameters

One of the most important benefits of direct detection using SPR biosensor technology is the determination of kinetics of molecular interactions. Reaction rate and equilibrium constants of interactions can be determined, e.g. the interaction

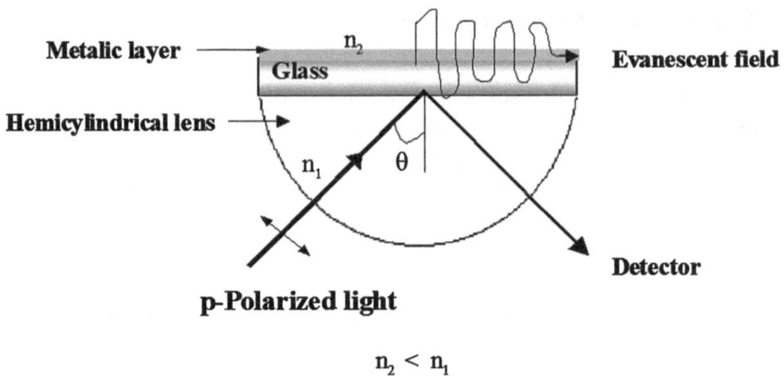

Fig. 7.1 Diagram of surface plasmon excitation. The bulk propagation vector, p-polarized light, increases $n1$ times inside a prism with a determined refractive index. By changing the angle of incidence, the component of the propagation vector parallel to surface can be matched to the surface plasmon wave vector on a metal surface (evanescent field) with a refractive index

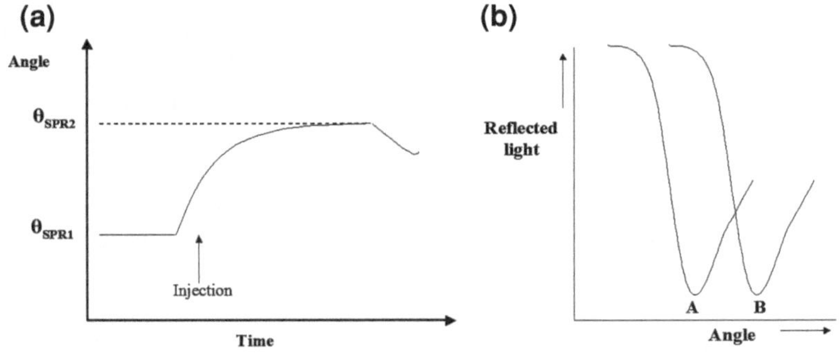

Fig. 7.2 **a** A sensorgram: the angle at which the dip is observed versus time. First, no change occurs at the sensor and a baseline is measured with the dip at SPR angle A. After injection of the sample (*arrow*) molecules will adsorb on the surface resulting in a change in refractive index and a shift of the SPR angle to position B. **b** The adsorption–desorption process can be followed in real time and the amount of adsorbed species can be determined by the change of refractive index

$A + B \rightarrow AB$ can be followed in real time with SPR technology, where A is the analyte and B is the ligand immobilized on the sensor surface. In addition, kinetic experiments can provide information on the thermodynamics, e.g. on the binding energy of processes. The typical range of the association and dissociation constant shows large variations and is dependent on, among other things, the temperature [18–20].

The use of SPR for the measurement of binding parameters, mainly in biological analysis, has been reported. These parameters include reaction kinetics (*k*a, *k*d), binding constants and determining the active concentration of molecules. When experiments are performed carefully, SPR biosensors can also be used to determine the binding stoichiometry and mechanism of the interaction [21–26].

Surface plasmon resonance is a very sensitive optical technique. However, the maximum sensitivity that can be obtained in several applications is limited by the amount of analyte or the molecular weight of those directly binded or adsorbed to the surface of the SPR substrate. Thus, the detection of small molecules can be carried out using a different strategy. Most often, small molecules are detected in a sandwich, competition or inhibition assay format [27–29].

SPR biosensors are sensing devices which consist of a biorecognition element that recognizes and is able to interact with a selected analyte and an SPR transducer, which translates the binding event into an output signal. The biorecognition elements are immobilized in the proximity of the surface of a metal film supporting a surface plasmon. In addition, biorecognition elements need to be immobilized on the sensor surface without affecting their biological activity. In principle, the molecules can be immobilized either on the surface or in a three-dimensional matrix [30, 31].

The change in the refractive index produced by the capture of biomolecules depends on the concentration of analyte molecules at the sensor surface and the properties of the molecules. Sensor response is proportional to the binding-induced refractive index change [32].

7.4 Surface Plasmon Resonance and Electrochemistry

All sensing techniques demonstrate specific strengths, yet sometimes overlapping, areas of application. In this case, both electrochemical and optical techniques, can allow analysis in real-time, in situ, non-destructive, label-free, thin films and interfaces analysis [33–35].

Electrochemical reaction occurring at the electrode surface is a heterogeneous process. Therefore, it is possible to detect the electrochemical process by using the SPR technique. SPR is sensitive to a range of processes taking place on or near a sensor chip. Thus, the combination of electrochemistry and SPR, the thin metal film on the substrate serves not only to excite surface plasmons, but also acts as a working electrode for electrochemical detection or control (Fig. 7.3). One advantage of the ESPR configuration is the ability to simultaneously obtain information about the electrochemical and optical properties of films with thicknesses in the nanometer range.

This interaction between SPR and electrochemistry can be relevant for important processes in the biological field, mainly for analyses that study the interactions with antigen–antibody, nucleic acids, cells, enzymes, micro-organisms, etc. [36–38].

Electrochemical-SPR is also a powerful tool for monitoring the build-up of complex interfacial architectures along with an in situ electrochemical characterization. For surface-attached biomembrane mimicks, SPR combined with electrochemical impedance spectroscopy is well established. ESPR has often been applied to study the formation and the properties of thin films and mono/multi-

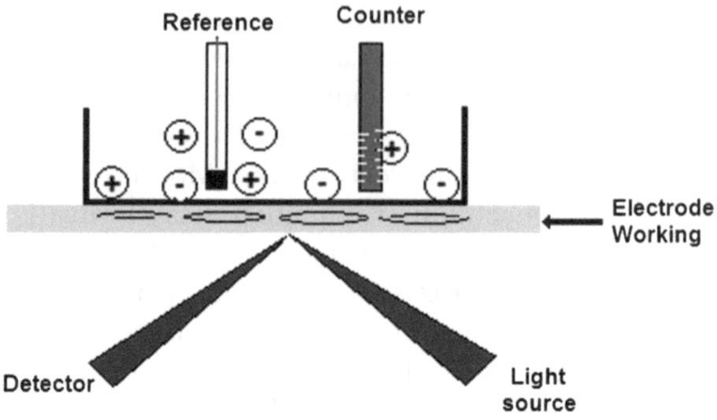

Fig. 7.3 Schema of the integration of electrochemistry and surface Plasmon resonance

biolayers using, for example, self-assembly or electro-polymerization methods and also to characterize ultra-thin film of conducting polymers and coupling principles to investigate electrochemical reactions with integrated optics and waveguide sensors [20, 39–47].

7.5 Bioanalytical Applications

Numerous detection strategies have also been developed for biosensing applications based on combining electrochemistry with SPR detection. Although most of the combined electrochemical and SPR studies utilized uniform electrode surfaces with traditional SPR detection, there have been several examples of combined electrochemical systems with SPR imaging, where the optical response of various locations on the electrode surface are investigated simultaneously. Simultaneous electrochemical and SPR analysis has been extensively used in the characterization of various conducting and electroactive polymer films to provide information about polymer assembly, redox transformations, electrochemically catalyzed processes and others applications [48, 49].

Moreover, the combination of SPR and an electrochemical allows for in situ kinetic investigation, a doping–dedoping process, and optical property changes during electropolymerization of electroactive monomers and has also been recently used in immunosensor applications [45, 49–52]

Gupta et al. [53] constructed a molecularly imprinted polymer (MIP) based on in situ electrochemical polymerization of 3-aminophenylboronicacid (3-APBA) on the bare gold chip for the detection of staphylococcal enterotoxin B (SEB), which is used as warfare agent. The control of the electropolymerization step was accomplished using SPR and cyclic voltammetry (CV) recorded simultaneously for 3-APBA with and without SEB (NIP). It was possible to calculate the SPR angle shift after polymerization and after SEB had been removed (MIP). The profile of CV in both cases was important to conclude the change in the sensor with biological molecule and after washing to remove. The MIP presented excellent sensitivity, with a detection limit of 0.05 fmol L^{-1} and good selectivity for similar toxins.

Also using a polymer as immobilization support, Dong et al. [54] reported the use of measurements of SPR and cyclic voltammetry simultaneously as detection systems for an immunosensor for the first time. The techniques were used to monitor the relationship between thickness of polymer film and growth of cycle number. Furthermore, it was possible to compare the immunosensor responses obtained by SPR and CV, as sensitivity and detection limit.

The formation and characterization of ultrathin film formed by poly(3-aminobenzoic acid) (PABA) was carried out by Sriwichai et al. [55] using ESPR for the development of immunosensor to detect human immunoglobulin G. With the aid of simultaneous measurements of SPR and CV it has been become possible to calculate the thickness and dielectric properties of a polymeric film, allowing that immunosensor responses can be related to its surface morphology. Another ESPR biosensor also based on PABA was developed by Baba et al. [56] to detect adrenaline. The polymer acts as a specific reaction site for adrenaline, presenting different electrochemical and SPR responses to those for uric and ascorbic acids, which are major interferences of the catecholamine studied. The two techniques were used to evaluate the electrodeposition of PABA and to obtain the calibration curves and the detection limit was set to 100 pmol L^{-1}.

The simultaneous measurements of SPR and electrochemical impedance were carried out using a flow injection analysis (FIA) cell by Bart et al. [5]. The FIA system was tested for interferon-γ detection using liposome bounded to the secondary antibody to increase the amount of mass for SPR detection. Liposome binding did not yield an impedance shift, but the different concentrations of interferon caused an increase in the impedance signal. In this way, it was possible to use the impedance-SPR measurements in this system. This immunosensor indicates the usefulness of the FIA cell for investigations of the ESPR method as a biosensor development.

To monitor the electrodeposition of ZnO film on a gold surface to prepare a glucose biosensor, Singh et al. [57] used an ESPR system. While the film was formed and accomplished using CV, the thickness was calculated by the SPR angle shift. This film was used to immobilize glucose oxidase via EDC/NHs activation of modified surface [6]. The work indicates promising applications of the system as a tool for studying bio-specific interactions and the development of others biosensors based on SPR detection.

The poly-o-phenylenediamine film and gold nanoparticles were combined to construct a biocompatible support for the immobilization of immunocompounds. The polymer film growth and the assembling of various sizes of gold nanoparticles were real-time monitored by SPR and electrochemical methods [4].

In this case, Xin et al. [58] applied scanning electrochemical microscopy (SECM) combined with SPR, SECM–SPR, to monitor in real-time the incorporation of Cu^{2+} by apo-metallothionein (apo-MT) immobilized on the SPR substrate and the release of Cu^{2+} from surface-confined metallothionein. The combination between these techniques allows detecting the structural and compositional changes on enzymes during their sequestration and release processes. The high sensitivity of the SPR instrument facilitates in situ measurements of infinitesimal changes in the structure of surface-confined protein molecules, at the same time as the SECM provides the versatility of controlling the local milieu that affects the protein property and function. The enhanced mass transfer rate at the SECM tip also improves the effect of limited mass transfer on the determination. It was possible to control with this coupled technique to control the extent of metal binding and also the binding stoichiometry and dynamics to be quantitatively determined. The same group also employed SECM-SPR for in situ monitoring of the incorporation of Hg^{2+} by apo-metallothionein immobilized on the SPR substrate. Hg^{2+} was anodically stripped from the Hg-coated SECM Pt tip and sequestered by apo-MT upon its diffusion to the SPR substrate. The high sensitivity of the SPR instrument enabled the detection of the changes in the composition and structure of apo-MT molecules that were induced by the metal sequestration of Hg^{2+}. It was possible to know the saturation co-ordination number of Hg^{2+} binding to apo-MT. The results observed by Xin et al. [59] are potentially useful for a deeper understanding of the detoxification mechanism of MT to mercury ion.

Schlereth (1999) used the SPR technique coupled with cyclic voltammetry to characterize monolayers of cytochrome-c and cytochrome-c-oxidase adsorbed on gold surfaces modified with different alkanethiol self-assembled monolayers [60]. Different behaviors for enzyme adsorption processes in the modified gold surface were observed. For modified mercapto propioni acid electrodes, the response observed for the cytochrome-c adsorbed may be explained as arising from a potential-dependent adsorption and for cytochrome-c-oxidase appears a conformational change between the two states of the adsorbed oxidase, which gives rise to two species with different electrochemical behaviour.

ESPR can be used in order to distinguish the enzyme activity of conducting polymer/glucose oxidase films, constructed by layer-by-layer processes, from changes in the film thickness and the dielectric constant. The results obtained indicated which doped state had the highest reflectivity change or was more sensitive for the optical signal Baba et al. [56]. This is not counterintuitive because the polypyrrole film is oxidized in the glucose sensing (reduction) event and thus the developed state shows the highest change, from a dedoped state to a more doped state [48, 61]. The results obtained also highlight the fact that the change of reflectivity can also be controlled by the doping state of the conducting polymer films.

Heaton et al. [62] used surface plasmon resonance spectroscopy to monitor hybridization kinetics for unlabeled DNA in tethered monolayer nucleic acid films on gold in the presence of an applied electrostatic field which can be used, in a reversible manner, to increase or decrease the rate of oligonucleotide hybridization.

The visualization of the electrochemical reaction distribution on structured and modified electrodes provides instant information about the relationship between electrochemical activity and physical structure. Iwasaki et al. [63] constructed an electron mediator type enzyme sensor using horseradish peroxidase on a gold electrode that also served as an SPR substrate. Thus, they used this substrate to perform the optical mapping of enzyme activity with electrochemical activation and controlled the electrochemical states of the mediator in cyclic voltammetry and imaged the degree to which the charged site density changed.

Wang et al. [4] used the electrochemical surface plasmon resonance method to investigate enzyme reactions in a bilayer lipid membrane based on immobilizing horseradish peroxidase in theses membrane lipids supported by the redox polyaniline. After each step of the detection of peroxide hydroxide carried out by peroxidase, the SPR sensor surface was completely regenerated by electrochemically reducing the oxidized polyaniline to its reduced state.

Although surface plasmon resonance-electrochemistry-based bioanalytical assays are most commonly associated with surface characterization, protein interaction analysis and drug discovery has gained increased interested. The main advantages of ESPR bioanalysis of SPR-based detection over alternative analytical techniques such as microbiological assays include ease of use, simpler and faster sample preparation and reduced assay time from days to minutes in some cases. SPR biosensors offer the clearest advantages in speed over alternative techniques that rely on biological readouts such as inhibition of microbial growth for detecting antibiotics.

7.6 Conclusion

This chapter describes a brief approach of some promising applications of surface plasmon resonance in the investigation of electrochemical processes in several bioanalytical applications with high sensitivity and data sampling in order to enable the ESPR as an excellent setting for research of interfacial processes in situ and in real time. From the above, it shows the promising nature of the combined use of surface plasmon resonance with electrochemical techniques, not only because of the sensitivity of the SPR technique, but also in view of the possibility of the future development of highly sensitive, highly specific, multi-analysis and nanoscale biosensors. Any advancement in this field will have an effect on the future of diagnostics, environmental and health care due to the range of opportunities it provides for a more complete study of the interface electrode-solution.

References

1. Situ, C., Buijs, J., Mooney, M.H., Elliott, C.T.: Trends Anal. Chem. **29**, 1305 (2010)
2. Wang, J., Munir, A., Li, Z., Zhou, H.S.: Biosens. Bioelectron. **25**, 124 (2009)
3. Mouri, R., Oishi, T., Torikai, K., Ujihara, S., Matsumori, N., Murata, M., Oshima, Y.: Bioorg. Med. Chem. Lett. **19**, 2824 (2009)
4. Wang, Q., Tang, H., Xie, Q., Jia, X., Zhang, Y., Tan, L., Yao, S.: Colloids Surf. B **63**, 254 (2008)

5. Bart, M., van Os, P.J.H.J., Kamp, B., Bult, A., Bennekom, W.P.: Sens. Actuators B. **84**, 129 (2002)
6. Szunerits, S., Rich, S.A., Coffinier, Y., Languille, M.A., Supiot, P., Boukherroub, R.: Electrochim. Acta **53**, 3910 (2008)
7. Wang, Y., Knoll, W.: Anal. Chim. Acta **558**, 150 (2006)
8. Schasfoort, R.B.M., Tudos, A.J. (eds.): Introduction to surface plasmon. In: Resonance Handbook of Surface Plasmon Resonance. The Royal Society of Chemistry, Cambridge (2008)
9. Englebienne, P. Hoonacker, A.V., Verhas, M.: Spectroscopy **17**, 255 (2003)
10. Homola, J., Yee, S.S., Gauglitz, G.: Sens. Actuators B. **54**, 3 (1999)
11. Kretschmann, E., Raether, H.Z.: Z. Naturforsch. **23A**, 2135 (1968)
12. Otto, A. Z.: Physik. **216**, 398 (1968)
13. Liedberg, B., Nylander, C., Lundstrom, I.: Biosens. Bioelectron. **10**, 1 (1995)
14. Raether, H.: Surface Plasmons on Smooth and Rough Surfaces and on Gratings. Springer, Berlin (1988)
15. Stegeman, G.I., Burke, J.J., Hall, D.G.: Opt. Lett. **8**, 383 (1983)
16. Burke, J.J., Stegeman, G.I., Tamir, T.: Phys. Rev. B. **33**, 5186 (1986)
17. Sarid, D.: Phys. Rev. Lett. **47**, 1927 (1981)
18. Nguyen, B., Tanious, F.A., Wilson, W.D.: Methods **42**, 150 (2007)
19. Day, Y.S.N., Baird, C.L., Rich, R.L., Myszka, D.G.: Protein Sci. **11**, 1017 (2002)
20. Peterlinz, K.A., Georgiadis, R.: Langmuir **12**, 4731 (1996)
21. Lackmann, M., Bucci, T., Mann, R.J., Kravets, L.A., Viney, E., Smith, F., Moritz, R.L., Carter, W., Simpson, R.J., Nicola, N.A.: Proc. Natl. Acad. Sci. U S A **93**, 2523 (1996)
22. Myszka, D.G.J.: Mol. Recognit. **12**, 390 (1999)
23. Rich, R.L., Myszka, D.G.: Curr. Opin. Biotech. **11**, 54 (2000)
24. Markgren, P.-O., Hämäläinen, M., Danielson, U.: Anal. Biochem. **265**, 340 (1999)
25. Morton, T.A., Myszka, D.G.: Methods Enzymol. **295**, 268 (1998)
26. Myszka, D.G., Jonsen, M.D., Graves, B.J.: Anal. Biochem. **265**, 326 (1998)
27. Matsubara, K., Kawata, S., Minami, S.: Appl. Spectrosc. **42**, 1375 (1988)
28. Zhang, L.M., Uttamchandan, D.: Electron. Lett. **24**, 1469 (1988)
29. Brockman, J.M., Nelson, B.P., Corn, R.M.: Annu. Rev. Phys. Chem. **51**, 41 (2000)
30. Mello, L.D., Ferreira, D.C.M., Kubota, L.T.: Enzymes as analytical tools in food processing. In: Enzymes in Food Processing: Fundamentals and Potential Applications. I K International Publishing House, New Delhi (2010)
31. Ferreira, D.C.M., Mello, L.D, Mendes, R.K., Kubota, L.T.: Biosensors for fruit and vegetable processing. In: Enzymes in Fruit and Vegetable Processing Chemistry and Engineering Applications. CRC Press, Boca Raton (2010)
32. Homola, J.: Chem. Rev. **108**, 462 (2008)
33. Arwin, H., Poksinski, M., Johansen, K.: Appl. Opt. **43**, 3028 (2004)
34. Elwing, H.: Biomaterials **19**, 397 (1998)
35. Goodall, D.G., Stevens, G.W., Beaglehole, D., Gee, M.L.: Langmuir **15**, 4579 (1999)
36. Eggins, B.: Chemical sensors and biosensors. In: Analytical Techniques in the Sciences. Wiley, West Sussex (2002)
37. Chaubey, A., Malhotra, B.D.: Biosens. Bioelectron. **17**, 441 (2002)
38. D'Orazio, P.: Clin. Chim. Acta **334**, 41 (2003)
39. Baba, A., Lübben, J., Tamada, K., Knoll, W.: Langmuir **19**, 9058 (2003)
40. Xia, C., Advincula, R., Baba, A., Knoll, W.: Langmuir **18**, 3555 (2002)
41. Schweiss, R., Lübben, J.F., Johannsmann, D., Knoll, W.: Electrochim. Acta **50**, 2849 (2005)
42. Damos, F.S., Luz, R.C.S., Kubota, L.T.: Electrochim. Acta **51**, 1304 (2006)
43. Abanulo, C., Harris, R.D., Sheridan, A.K., Wilkinson, J.S., Bartlett, P.N.: Farad. Disc. **121**, 139 (2002)
44. Bradshaw, J.T., Mendes, S.B., Armstrong, N.R., Saavedra, S.S.: Anal. Chem. **75**, 1080 (2003)
45. Badia, A., Arnold, S., Scheumann, V., Zizlsperger, M., Mack, J., Jung, G., Knoll, W.: Sens. Actuators B. **54**, 145 (1999)

46. Lavers, C.R., Harris, R.D., Hao, S., Wilkinson, J.S., Odwyer, K., Brust, M., Schiffrin, D.J.J.: Electroanal. Chem. **387**, 11 (1995)
47. Willner, I., Katz, E.: Angew. Chem. Intl. Ed. **43**, 6042 (2004)
48. Baba, A., Knoll, W., Advincula, R.: Rev. Sci. Instrum. **77**, 064101 (2006)
49. Sriwichai, S., Baba, A., Deng, S.X., Huang, C.Y., Phanichphant, S., Advincula, R.C.: Langmuir **24**, 9017 (2008)
50. Knoll, W.: Annu. Rev. Phys. Chem. **49**, 569 (1998)
51. Baba, A., Park, M.K., Advincula, R.C., Knoll, W.: Langmuir **18**, 4648 (2002)
52. Gouzy, M.-F., Keß, M., Krämer, P.M.: Biosens. Bioelectron. **24**, 1563 (2009)
53. Gupta, G., Singh, P.K., Boopathi, M., Kamboj, D.V., Singh, B., Vijayaraghavan, R.: Thin Solid Films **519**, 1115 (2010)
54. Dong, H., Cao, X., Li, C.M., Hu, W.: Biosens. Bioelectron. **23**, 1055 (2008)
55. Sriwichai, S., Baba, A., Phanichphant, S., Shinbo, K., Kato, K.: Sens. Actuators B. **147**, 322 (2010)
56. Baba, A., Mannen, T., Ohdaira, Y., Shinbo, K., Kato, K., Kaneko, F., Fukuda, N., Ushijima, H.: Langmuir **26**, 18476 (2010)
57. Singh, N.K., Jain, B., Annapoorni, S.: Sens. Actuators B Chem. **156**, 383 (2011)
58. Xin, Y., Gao, Y., Guo, J., Chen, Q., Xiang, J., Zhou, F.: Biosens. Bioelectron. **24**, 369 (2008)
59. Xin, N., Xin, Y., Gao, Y., Xiang, J.: Microchim Acta (2011). doi:10.1007/s00604-011-0598-z
60. Schlereth, D.D.J.: Electroanal. Chem. **464**, 98 (1999)
61. Calvo, E.J., Forzani, E., Tero, M.J.: Electroanal. Chem. **231**, 538 (2002)
62. Heaton, R.J., A.W., Peterson, Georgiadis, R.M.: PNAS **98**, 3701 (2001)
63. Iwasaki, Y., Tobita, T., Kurihara, K., Horiuchi, T., Suzuki, K., Niwa, O.: Biosens. Bioelectron. **17**, 783 (2002)